Scalable Multicasting over Next-Generation Internet

T0181766

Xiaohua Tian • Yu Cheng

Scalable Multicasting over Next-Generation Internet

Design, Analysis and Applications

 Springer

Xiaohua Tian
Department of Electronic Engineering
Shanghai Jiao Tong University
800 Dongchuan Road
SEIEE Buildings 5-301
Shanghai
China, People's Republic

Yu Cheng
Department of Electrical and Computer
Engineering
Illinois Institute of Technology
3301 S Dearborn
Chicago, IL
USA

ISBN 978-1-4899-9527-8   ISBN 978-1-4614-0152-0 (eBook)
DOI 10.1007/978-1-4614-0152-0
Springer New York Heidelberg Dordrecht London

To my parents, for your love, support, and understanding.
Xiaohua Tian

To my wife, Yanning and our daughter, Annabelle.
Yu Cheng

Preface

Next generation Internet faces two significant changes. First, a variety of network multimedia applications have sprung up and infiltrated into every aspect of people's lives, which makes it the basic functionality of the next-generation Internet to provide high quality multimedia services, such as online gaming and video, IPTV, and video conferencing. Second, the rapid advances in microelectronics technology have offered new opportunities to the development of next-generation Internet by providing networking devices with faster computing capability, capacious storage space, and high performance price ratio. In recent decade, how to take the advantage of the opportunity by hardware technology to improve the overall performance of the next-generation Internet for providing users better multimedia experience has been a research hotspot in the computer communication community.

This book studies a critical research issue in this area, that is, how to exploit the enhanced hardware capacity in next-generation Internet to design a scalable multicast protocol for supporting large-scale multimedia services. Specifically, we will investigate the following issues. How the multicast mechanism is working? What is the state of the art in multicast protocol design? Why do we need to design a multicast protocol in the context of next-generation Internet? How do we perform a theoretical analysis on multicast protocols? How do we apply a new protocol design to practical networking applications? How dowe implement a new protocol in the ns-2, a network simulator well known as hard-to-extend?

The book will be of interest to researchers in the area of future Internet, application-oriented networking, and multimedia communication. It is also directed toward senior undergraduate and graduate students who major in computer networks.

Based on our continuous research on the topic in the past 5 years, we present the development of a scalable multicast protocol for next-generation Internet in a systematic manner. Chapter 1 gives an introduction to the issue of the scalable multicast protocol over next-generation Internet. We describe three aspects of the issue: first, design a scalable multicast with an application-oriented networking (AON) approach in the context of the next-generation Internet; second, integrate the proposed application-oriented multicast (AOM) scheme into IPTV system seamlessly and improve the performance of IPTV service; third, develop a generic simulation framework not only to implement the multicast scheme but also to facilitate the future research on next-generation Internet.

Chapter 2 presents a comprehensive survey on the multicast protocol design in the context of Internet evolution in the past 2 decades. We describe the advancement of the IP multicast mechanisms, categorize the methodologies of overlay multicast protocols, and introduce modern multicast schemes proposed in recent decade. After that, we discuss the technique tendency of next-generation Internet with its influence to multicast protocol design. The typical IPTV network architecture is described as an example of applying multicast to multimedia applications, with the channel-zapping issue of IPTV over IP multicast explained in detail.

Chapters 3 and 4 focus on the design of a scalable multicast protocol specifically for the next-generation Internet with more intelligent networking nodes. Chapter 3 proposes a scalable AOM multicast service model and identifies five important practical design issues. The first two design issues are resolved in Chap. 3, with theoretical analysis and simulation results provided to evaluate the performance of the proposed scheme. Chapter 4 resolves the rest of practical design issues identified in Chap. 3. The elaborated design enables the AOM scheme to better accommodate to the next-generation Internet environment and further improves the performance of AOM protocol, which is validated by comprehensive theoretical analysis and simulation results.

Chapter 5 investigates the application of AOM protocol to IPTV service, with the focus on reducing the channel-zapping time for IPTV systems. We first analyze existing solutions for zapping acceleration and find that the time-shifted subchannel (TSS)-based solution has remarkable advantages over others. A systematic analysis of the TSS-based model is developed to study its fundamental properties from a theoretical perspective. Then, an AOM-Assisted Zapping Acceleration (AAZA) scheme is proposed to realize the TSS-based model. The advantageous properties of AOM can improve the robustness and flexibility of IPTV system, compared with IP multicast-based implementation. AAZA is realized in ns-2, and a MPEG-4 video stream is multicast over a practical network topology.

Chapter 6 develops a generic AON (GAON) simulation framework compatible with ns-2 to facilitate the research on next-generation Internet with the AON approach. With GAON, developers can conveniently enhance the ns-2 nodes with customized functionalities and seamlessly incorporate them into the regular ns-2 system. GAON provides a generic scenario control interface, through which ns-2 users can flexibly load/unload customized GAON-processing agents on network nodes. The regular ns-2 node structure is extended, where a GAON agent classifier is set upto dispatch GAON traffic to correct GAON agents. Moreover, a unified interface to the ns-2 built-in routing table is developed to facilitate GAON agents in forwarding packets.

Chapter 7 gives possible open research issues in the future.

Shanghai   Xiaohua Tian
Chicago   Yu Cheng

Acknowledgements

Xiaohua and Yu want to thank the entire Springer Science+Business team, which has provided us a lot of help in completing this book. Our special thanks go to Brett Kurzman and Elizabeth Dougherty, for their management, patience, and encouragement.

Contents

Acronyms

AAD	Average access delay
AAZA	Application oriented multicast assisted zapping acceleration
ADC	Application oriented multicast destinations cache
AOM	Application oriented multicast
AON	Application oriented networking
APR	Average packets reception
AS	Autonomous system
BFJ	BGP-view based fast-join
BGMP	Border gateway multicast protocol
BGP	Border gateway protocol
BRA_BF	The bloom filter encoding the subset of receivers' destinations in DST_BF after the TBR/SDR processing in DOM
CAN	Content addressable networks
CBT	Core based tree
CDF	Cumulative distribution function
DC	Data center
DHCP	Dynamic host configuration protocol
DHT	Distributed hash table
DM	Dense mode
DNS	Domain name system
DR	Designated router
DST_BF	The bloom filter encoding receivers' destinations, which is attached to the DOM multicast packet
DVMRP	Distance vector multicast routing protocol
FID	First I-frame delay
FRM	Free riding multicast
GAON	Generic application oriented networking
GHL	Group hosts list
GID	Group identifier
GOP	Group of pictures
GRP_BF	Group bloom filter

IGMP	Internet group management protocol
IPTV	Internet protocol television
IPv6	Internet protocol version 6
IRDR_BF	The interface RDR bloom filter at TBR/SDR, installed by MUM message and encoding the prefix associated with the subscribing RDR
LIT	Link identifier tag
MAD	Multicast with adaptive dual-state
MAZA	Multicast assisted zapping acceleration
MBGP	Multiprotocol extensions to border gateway protocol-4
mCH	Main channel
MGL	Multicast group list
MICC	Multicast instant channel change
MOSPF	Multicast open shortest path first
MPEG	Moving picture experts group
MSDP	Multicast source discovery protocol
MT	Membership tree
MUM	Membership updating message
OSPF	Open shortest path first
OTCL	Object oriented tool command language
PAAD	Per-node average access delay
PIM	Protocol independent multicast
PPR	Per-node packet reception
QoE	Quality of experience
RDR	Designated border router of receiver-side domain; when multicast routing/forwarding is considered, it also represents the prefix associated with the receiver domain. The meaning of RDR will be clear in the context
REUNITE	Recursive unicast tree
RP	Rendezvous point
RPF	Reverse path forwarding
SA	Source active
sCH	Sub-channel
SDR	Designated border router of source-side domain
SM	Sparse mode
SOA	Service oriented architecture
SP	Service provider
SPT	Shortest path tree
SRC	Data source
SSM	Source specific multicast
SST	Sub-channel states table
STB	Set-top box
TBR	Border router of transit domain
TC	Traffic class
TCL	Tool command language

TOS	Type of service
TSS	Time-shifted sub-channels
TTL	Time-to-live
VoD	Video-on-demand
Xcast	Explicit multicast
XML	Extensible markup language

Chapter 1
Introduction

Abstract This chapter gives the big picture of this book. Section 1.1 introduces the issue of multicast. Multicast is a kind of communication paradigm, where data is delivered to multiple receivers. Efficient multicast requires a set of mechanisms specifically designed for the purpose. Existing solutions of multicast are either poor scalable or bandwidth inefficient, which are briefly discussed in Sect. 1.2. A scalable and efficient multicast mechanism is still an open technique issue. Section 1.3 presents the motivation and objectives of this book. This book aims at proposing a scalable and efficient multicast protocol for next-generation Internet, where the network routers are enhanced with application-level intelligence. The proposed protocol will be applied to the IPTV service. Moreover, a generic application-oriented networking simulation framework in Network Simulator ns-2 will be developed to facilitate the research of next-generation Internet.

1.1 Unicast, Broadcast, and Multicast

The next-generation Internet will accommodate various multimedia applications over a common IP-based transport infrastructure. Many real-time multimedia applications require data to be delivered from a sender to multiple receivers. Examples of such applications include IPTV, online multiplayer games, video conferencing, etc. The communication paradigm where data is delivered from a sender to multiple receivers is called *multicast*.

A feasible way to implement multicast is for the sending node to send a separate copy of the packet to each receiver, as shown in Fig. 1.1a. Given the two receivers, the source node simply generates two copies of the packet, with each delivered to one of the receivers with unicast. However, this approach incurs redundant traffic as the same packet is transmitted over the same link multiple times. Another possible solution is to broadcast the packet to all destinations as illustrated in Fig. 1.1b, which also leads to redundant traffic to nodes that are not interested in the packet.

X. Tian and Y. Cheng, *Scalable Multicasting over Next-Generation Internet: Design, Analysis and Applications*, DOI 10.1007/978-1-4614-0152-0_1,
© Springer Science+Business Media New York 2013

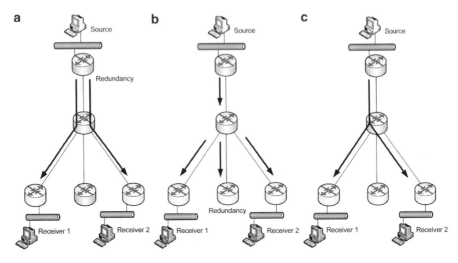

Fig. 1.1 Unicast, broadcast, and multicast

The drawback of inefficiency in the two solutions is significant when sending data to a large number of receivers. The ideal multicast delivery is illustrated in Fig. 1.1c; it requires a set of mechanisms which are specifically designed for the purpose.

1.2 Scalability and Bandwidth Efficiency

Many protocols have been proposed since Deering's pathbreaking work [14] was published in 1990s. The traditional multicast solutions are implemented at the network layer, where the IP routers need to communicate with each other to construct and maintain a tree structure according to a distributed multicast routing algorithm [4, 7, 18, 20, 31]. The IP multicast is bandwidth efficient in data delivery but poor scalable in managing the multicast tree [1, 17], since each router needs to maintain the multicast forwarding states for every group passing through. Although various multicast protocols, for example, *dense mode* protocols [27, 29, 31], *sparse mode* protocols [4, 20], and inter-domain protocols [1, 39], have been proposed to reduce the messaging overhead and the amount of states at each router for enabling a single group, the messaging overhead and memory cost grow linearly with the number of multicast groups being supported by the router. The fundamental reason is that the members associated with a group are usually not topologically concentrated, and forwarding states are not easily aggregateable [9, 16, 17, 38]. Due to the scalability issue, implementation complexity, and some other technical and marketing reasons [17, 21, 25], the IP multicast has not been widely deployed.

In most of the existing multicast protocols, an active group is uniquely identified by an IP address, that is, a class-D address according to IPv4. Such a global address allocation scheme is not scalable and flexible when the number of active groups grow significantly. The global address allocation implies that the total number of active multicast groups is limited by the class-D address space. Even if the address space may not be a problem under IPv6 [13], hooking up the group ID with routing is very inflexible. For example, if a service provider (SP) tends to expand its channel list or adjust the multicast addresses allocated to some channels, all the routing information with the network has to be reestablished. An interesting idea of localized group address allocation was proposed in [28, 32], where the combination of the source node address and a local group ID serves as a unique group ID. However, the benefit of localized address allocation can be fully exploited only when group ID is decoupled with multicast routing.

The more recent overlay multicast establishes the data-dissemination structure at the application layer [35, 36], where trees or other delivery structures are constructed at the application layer [17, 35, 36]. Each overlay link in the structure is an end-to-end unicast path between two hosts. Overlay multicast is convenient for deployment, since the underlying unicast infrastructure needs no modification. Nevertheless, overlay multicast performs less efficient than IP multicast in bandwidth utilization, because it induces redundant traffic at the network layer [5, 17]: it is not a rare case that separate overlay links pass through the common physical links in the underlying transport network.

1.3 Multicasting Over Next-Generation Internet

1.3.1 Multicast with an AON Approach

In this book, we reinvestigate the multicast issue by examining the fundamental evolution trend to the next-generation Internet: (1) many academic and industrial efforts to improve the Internet functionalities, especially the evolution of multicast protocols, have shown the tendency to incorporate more application-layer intelligence into the network layer; (2) improvements in hardware computation power and cost efficiency facilitate the networking equipments to execute more complex software and assume more computation burden. For multicast, we note that while the overlay approach resorts to application-layer intelligence, the IP-based solution tilts the processing to the network layer. The long-lasting issue that neither the network-layer nor the application-layer approach itself can achieve a generic scalable multicast solution reveals that multicasting by nature requires incorporating intelligence from both application and network layer. That is, identifying the users associated with a multicast group requires application-layer membership management, while delivering data to the proper destinations needs network-layer support according to the application-layer membership information [42].

In fact, incorporating the application-layer intelligence at the network layer has become a mainstream idea to design the modern multicast protocols. REUNITE [28] stores the destination addresses of a multicast group at the network routers located at the intersection points along a multicast tree to facilitate multicast forwarding, so that the forwarding table size in these routers can be significantly reduced; however, it still needs to maintain group-specific information through a soft-state mechanism, which still incurs considerable memory and message overheads. In Xcast [8], each packet carries a list of destination addresses and the network router can process the destinations list for multicasting. Although Xcast takes the similar service model as adopted in this article, it is not scalable, because the limited size of the Xcast header constrains the protocol only applicable in the scenario with a small number of multicast receivers. FRM scheme [15], most closely related to our study, encodes the multicasting tree into the packet using a bloom filter and computes packet copies and output interfaces by checking the router's neighboring edges against the embedded tree. Our method instead only encodes receiver destinations in the packet, which can reduce bandwidth overhead and accelerate the group joining process. The LIPSIN scheme [34] takes the similar methodology as FRM with enhanced techniques to reduce false positive incurred by the bloom filter processing, but it requires the protocol stack and underlying network architecture to be fundamentally changed. In MAD [66], it is proposed to adaptively switch between IP multicast mode and overlay multicast mode, depending on the service scenarios. Such a scheme can mitigate memory overhead when there are a large number of groups with infrequent data traffic, but cannot fundamentally solve the scalability issue.

In this book, we interpret the idea of incorporating application-level intelligence into the network as *application-oriented networking* (AON).[1] In fact, enhancing network nodes with application intelligence has become one of the mainstream ideas to design the next-generation Internet [10, 37], stimulated by various applications, including firewalls, Web proxies/caches, mobile gateways, service-oriented architectures, in addition to the multicast. Cisco has already started to produce network devices with application intelligence to enhance the deployment, management, and integration of network applications [2, 19]. However, as far as we know, all of the current AON[2] studies are directed to facilitate the upper-layer applications [11, 12, 37]. Our perspective is that AON also provides a methodology to streamline the design of networking functionalities, with the focus of this research on multicast. We term the objective multicast scheme as application-oriented multicast (AOM), as it will be designed with the concept of AON.

---

[1] The abbreviation *AON* stands for either application-oriented networking or application-oriented network, depending on the context where it is used.

[2] In this research, we enrich the denotation of AON compared to Cisco's definition: AON represents any new applications, services, or networking protocols that exploit the application intelligence built into the network; the researches in the AON context are termed as *application-oriented* studies.

1.3.2 Application of AOM to IPTV

IPTV is an increasingly popular Internet service in recent years. In IPTV system, only requested media data are forwarded to subscribers as a multicast stream to save bandwidth, which is quite different from the "broadcasting-for-all" paradigm in the traditional TV transmission. Currently, the transport technique supporting IPTV is IP multicast. In this book, we want to find if the performance of IPTV service can be improved by restructuring the underlying multicast mechanism. Our main focus is the channel zapping time, an important quality of experience (QoE) metric for IPTV service. Recently, IPTV zapping acceleration scheme based on time-shifted subchannels (TSS) is introduced [6]. The TSS-based model is more advantageous than most of existing channel acceleration schemes based on the auxiliary stream; however, the fundamental properties and performance of TSS-based model are still not well understood from a theoretical perspective. Moreover, the TSS-based model implemented with IP multicast will incur frequent messaging overhead and will be easily affected by the joining message signaling time in practice. This book will develop a systematical analysis for the TSS-based model and reveal some of its important properties from a theoretical perspective. Then, we will compare the performance of TSS-based model implemented with IP multicast and the proposed AOM, respectively.

1.3.3 Developing a Generic AON Simulation Framework

This book resorts to the Network Simulator ns-2 to measure the performance of the proposed mechanisms. ns-2 is the most widely accepted simulator in the networking community, but it is also well known that ns-2 lacks friendly interface for developers. We intend to develop a generic AON technique simulation platform compatible with ns-2, so that it not only helps shed light on the veiled ns-2 mechanisms but also facilitates the AON-related research in the future. To our best knowledge, the existing literature lacks the practical instructions on how to enhance the ns-2 nodes with customized intelligence. The ns-2 manual [41] is mainly written as a reference book, which provides limited information on how to extend ns-2 modules. Teerawat and Ekram's book [23] presents the fundamental principles of ns-2, but mentions little about how to extend the ns-2 node structure. The ns-2 extension examples available online [22, 30, 43] or in the technical papers [3, 24] are usually implemented in an ad hoc manner, with limited value of reference to develop a generic simulation platform. Some simulation frameworks have been proposed to facilitate the ns-2 extension, where systematical methods are provided; however, these frameworks either require to replace the ns-2 built-in core, for example, GenMCast [19], or introduce other software technique [26], which makes the complexity of ns-2 even more intractable. Thus, developing a generic AON simulation framework has much practical meaning, which saves researchers, especially young researchers, much time and energy to manipulate the simulator.

1.3.4 Objectives

The objectives of this book are to develop an efficient and scalable multicast mechanism for the next-generation Internet, apply the proposed multicast scheme to IPTV system, and develop a generic simulation framework to facilitate the AON research in the future. While the existing unicast mechanism is appropriately exploited, new multicast membership scheme and forwarding protocol are proposed. To integrate the AOM into the legacy infrastructure, the incremental deployment solution is also investigated. Moreover, application of the proposed multicast techniques to IPTV service is studied. In order to facilitate the future research related to the AON concept, a generic AON simulation platform is to be developed and integrated into ns-2 simulator [40]. Specifically, this book will:

- Comprehensively survey different multicast mechanisms in the literature.
- Propose a new multicast protocol AOM in the context of next-generation Internet, which is scalable in routing, forwarding, and address allocation.
- Extend the AOM protocol to support inter-domain environment, where the Internet features including the longest-prefix matching, route aggregation, and asymmetric inter-domain routing are present.
- Develop an efficient membership updating scheme so that the real-time requirement of some applications could be satisfied.
- Develop an incremental deployment solution of AOM for next-generation Internet, so that it can work in a network, where only a small fraction of the routers have AOM-aware intelligence while others are legacy routers.
- Apply the proposed AOM protocol to IPTV service, which seeks to utilize the advantageous properties of AOM to improve the performance of IPTV systems.
- Provide numerical, theoretical analysis, and computer simulation results on the performance of the proposed AOM protocol.
- Develop a generic AON simulation platform as a package for ns-2 simulator to facilitate the AON research in the community.

References

1. Almeroth KC (2000) The evolution of multicast: from the mbone to interdomain multicast to internet2 deployment. IEEE Network 14:10–20
2. Anthias T, Sankar K (2006) The network's new role. http://acmqueue.com/modules.php?name=Contentpa=showpagepid=391. Accessed May 2006
3. Ballantyne R, Feng T, Trajkovic L (2004) Implementation of BGP in a network simulator. In: Proc. Applied Telecommunications Syposium, Apr 2004, pp 149–154
4. Ballardie A (1997) Core based trees (CBT) multicast routing architecture. IETF RFC 2201
5. Bauer, F, Varma A (1995) Degree-constrained multicasting in point-to-point networks. In: Proc. IEEE INFOCOM, Mar 1995
6. Bejerano, Y, Koppol PV (2009) Improving zapping response time for IPTV. In: Proc. IEEE INFOCOM, Mar 2009, pp 1971–1979
7. Bhattacharyya S (2003) An overview of source-specific multicast (ssm). IETF RFC 3569

8. Boivie R, Feldman N (2001) Explicit multicast (Xcast) basic specification. Internet Draft (2001)

9. Briscoe R, Tatham M (1997) End to end aggregation of multicast addresses. Internet Draft (1997)

10. Clark D New arch: future generation Internet architecture. Technical Report. http://www.isi.edu/newarch/iDOCS/final.finalreport.pdf

11. Cisco Systems, Cisco application-oriented networking facilitates intelligent radio frequency identification processing at edge. http://www.cisco.com/en/US/products/ps6480/prod_brochure0900aecd8032811f.html

12. Cisco Systems, Cisco application-oriented networking streamlines finacial market-data and trade-order latency. http://www.cisco.com/en/US/products/ps6480/prod_brochure0900aecd804b0abe.html

13. Deering S (1998) Internet protocol, version 6 (IPv6) specification. IETF RFC 2460

14. Deering S, Cheriton D (1990) Multicast routing in datagram internetworks and extended lans. ACM Trans Comp Syst 8:85–110

15. Ermolinskiy A, Ratnasamy S, Shenker S (2006) Revisiting IP multicast. In: Proc. ACM SIGCOMM, 2006, pp 15–26

16. Estrin D, Radoslavov P, Govindan R (1999) Exploiting the bandwidth-memory tradeoff in multicast state aggregation. Technical report. University of Southern California

17. Fahmy S, Kwon M (2007) Characterizing overlay multicast networks and their costs. IEEE/ACM Trans Networking 15:373–386

18. Faranacci D, Jacobson V, Liu C, Deering S, Estrin D, Wei L (1996) The pim architecture for wide area multicasting. IEEE/ACM Trans Networking 4:153–162

19. Genmcast – generic multicast extension for ns-2. www.cse.nd.edu/~striegel/research/GenMCast/index.html

20. Helmy A, Thaler D, Deering S, Handley M, Wei L, Estrin D, Farinacci D (1998) Protocol independent multicast-sparse mode (pim-sm): protocol specification. IETF RFC 2362

21. Holbrook HW, Cheriton DR (1996) IP multicast channels: express support for single-source multicast applications. In: Proc. ACM SIGCOMM, Aug 1999, pp 65–78

22. Implementing a new nanet unicast routing protocol in ns2. http://masimum.inf.um.es/nsrt-howto/pdf/nsrt-howto.pdf

23. Issariyakul T, Hossain E (2009) Introduction to network simulator ns2. Springer, New York

24. Kong R (2008) The simulation for network mobility based on ns2. In: Proc. International Conference on Computer Science and Software Engineering, 2008, pp 1070–1074

25. Lyles B, Kassem H, Diot C, Levine B, Balensiefen D (2000) Deployment issues for IP multicast service and architecture. IEEE Network 14:78–88

26. Miozzo M, Rossi M, Baldo N, Maguolo F, Zorzi M (2007) ns2-miracle: a modular framework for multi-technology and cross-layer support in network simulator 2. In: Proc. ValueTools'07, Oct 2007

27. Moy J (1994) Multicasting extentions to OSPF. IETF RFC 1584

28. Ng TSE, Stoica I, Zhang H (2000) Reunite: a recursive unicast approach to multicast. In: Proc. IEEE INFOCOM, Mar 2000, pp 1644–1653

29. Nicholas J, Adams A, Siadak W (1998) Protocol independent multicast-dense mode (pim-dm): protocol specification (revised). IETF RFC 3973

30. NS2 learning guide. http://140.116.72.80/~smallko/ns2/ns2.htm

31. Partridge C, Waitzman D, Deering S (1988) Distance vector multicasting routing protocol. IETF RFC 1075

32. Perlman R (1999) Simple multicast: a design for simple, low-overhead multicast. Internet Draft

33. Ramakrishnan KK, Srivastava D, Cho TW, Rabinovich M, Zhang Y (2009) Enabling content dissemination using efficient and scalable multicast. In: Proc. IEEE INFOCOM, Apr 2009, pp 1980–1988

34. Rothenberg CE, Arianfar S, Jokela P, Zahemszky A, Nikander P (2009) LIPSIN: line speed publish/subscribe inter-networking. In: Proc. ACM SIGCOMM, 2009, pp 195–205

35. Seshan S, Chu Y, Rao S, Zhang H (2001) Enabling conferencing applications on the Internet using an overlay multicast architecture. In: Proc. ACM SIGCOMM, Aug 2001, pp 55–67
36. Shi SY, Turner JS (2002) Multicast routing and bandwidth dimensioning in overlay networks. IEEE J Select Areas Commun 20:1444–1455
37. Tennenhouse DL, Wetherall DJ (2002) Towards an active netwrok architecture. In: Proc. DARPA Active Networks Conference and Exposition, 2002, pp 2–15
38. Thaler D, Handley M (2000) On the aggregatble of multicast forwarding state. In: Proc. IEEE INFOCOM, Mar 2000, pp 1654–1663
39. Thaler D, Alaettinoglu C, Estrin D, Kumar S, Radoslavov P, Handley M (2000) The MASC/BGMP architecture for inter-domain multicast routing. IEEE Network 14:10–20
40. The network simulator – ns-2. http://www.isi.edu/nsnam/ns
41. The network simulator – ns-2: Documentation. http://www.isi.edu/nsnam/ns
42. Tian X, Cheng Y, Liu B (2009) Design of a scalable multicast scheme with an application-network cross-layer approach. IEEE Trans Multimedia 11:1160–1169
43. Tutorial for the network simulator "ns". http://www.isi.edu/nsnam/ns/tutorial/

Chapter 2
Background and Literature Review

Abstract This chapter explains related works to this book in detail. Section 2.1 reviews the evolution of IP multicast. The working principles of classic intra- and inter-domain IP multicast mechanisms are described. IP multicast has the scalability issue, which stimulates the research on overlay multicast. Section 2.2 presents main approaches to implement overlay multicast. Overlay multicast is implemented at the application layer, but the network layer knowledge is still needed for bandwidth efficiency. Section 2.3 introduces some representative multicast mechanisms proposed recently. The feature they have in common is to enhance network routers with more intelligence. This methodology is termed as application-oriented networking (AON). Section 2.4 explains the concept of AON in detail. Section 2.5 introduces an application of multicast mechanism, IPTV. The IPTV definition, architecture, and a specific technique issue, channel zapping, are discussed. This chapter is summarized in Sect. 2.6.

2.1 The Evolution of IP Multicast

2.1.1 Intra-Domain Multicast

The initial multicast research efforts focused on a single flat network topology, which is referred to as intra-domain multicast. Since the first Internet-wide experiments of multicast in 1992 [26], leveraging the unicast routing information has been the cornerstone idea underpinning various multicast protocols. There have been three basic approaches to exploit the unicast routing for multicast as following.

Source-Rooted Shortest-Path Tree. The first is to use *reverse path forwarding* (RPF) [25] against unicast table to construct a source-rooted shortest-path tree (SPT) in a *broadcast-and-prune* manner. The distance vector multicast routing protocol (DVMRP) [63] and Protocol independent multicast-dense model (PIM-DM) [61] had been developed based on this principle, although some implementation details

X. Tian and Y. Cheng, *Scalable Multicasting over Next-Generation Internet: Design,*
Analysis and Applications, DOI 10.1007/978-1-4614-0152-0_2,
© Springer Science+Business Media New York 2013

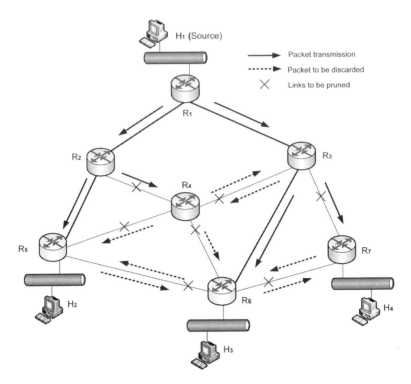

Fig. 2.1 Broadcast-and-prune

are different [4]. Figure 2.1 illustrates how the multicast packets from H_1 are forwarded with this approach:

- The source H_1 first broadcasts each packet on its local network, and the attached designated router R_1 receives the packet and forwards the packet copies on all output interfaces.
- Each router receives the packet copy performs a RPF check, that is, check the unicast routing table to see if the incoming interface on which the packet is received is the interface the router could use as output interface to reach the source. All packets received on the proper interface are further duplicated and forwarded on all output interfaces, and others are discarded. For example, R_3 forwards the packet from R_1 but discards the packets from R_4.
- With the help of the internet group management protocol (IGMP), the designated router of end hosts (e.g., R_5 and R_6) can discover if there are existing members for a given group. When the multicast packet for the group arrives, it forwards the packet to the subnet.
- The second phase of the broadcast-and-prune approach ensures packets are delivered to a router only if the router is on a path to a multicast group member. To this end, when a router receives a multicast packet on an interface that

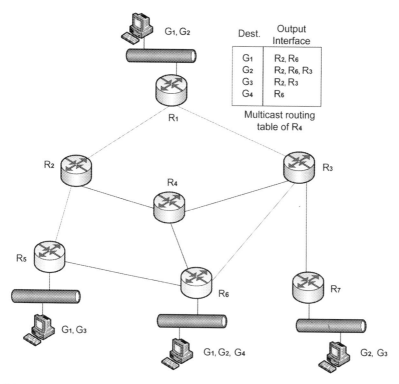

Fig. 2.2 MOSPF

is not the RPF interface, the router sends a prune message on that interface. Moreover, when a router finds itself not connected to a subnet with group members, it sends a prune message on its RPF interface. Figure 2.1 shows the links pruned according to the rules and highlights the source-rooted SPT eventually constructed.

Multicast Extensions to OSPF. The second approach is utilizing mutlicast extensions to OSPF (MOSPF) [59] to exploit open shortest path first (OSPF) to flood an OSPF area with the group membership information so that each MOSPF router can independently construct the shortest-path tree for each source and group. The shortest-path tree for a group defines the output interfaces for each subnet that has members for that group. Figure 2.2 shows an example of OSPF domains and the shortest-path tree in R_4, where the next-hop nodes are used to represent corresponding output interfaces for demonstration convenience.

Core/Rendezvous-Based Tree. In the third approach, each receiver can send explicit joining message to a selected core or a rendezvous point (RP), along the unicast shortest path between them, to construct a bidirectional or unidirectional shared tree, which results in the core based trees (CBT) protocol [5] and the Protocol independent multicast-sparse model (PIM-SM) [36], respectively. We use PIM-SM

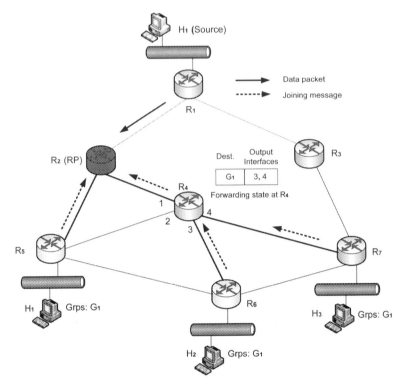

Fig. 2.3 PIM-SM

as our example to briefly describe the approach in this category, as PIM-SM is much more widely used than CBT. Figure 2.3 illustrates the PIM-SM protocol operations:

- A RP must be preconfigured. Different groups may use different routers for RPs, but a group can only associated with a single RP. The RP selection could be done by some bootstrap protocol [4].
- Receivers send explicit *joining messages* to the RP. The initial joining message creates a forwarding state in each router along the path from the receiver to the RP (e.g., the forwarding state at R_4 in Fig. 2.3). The subsequent joining message modifies the forwarding state in its first-passed router that has been already on the multicasting tree by adding corresponding output interfaces. For example, when the joining message from H_3 arrives at R_4, the state $(G_1\ 3)$ set up by H_2 has been there, R_4 stops forwarding the joining message to the RP, it just adds interface 4 to the state instead. By continuous observing the joining messages sent from different receivers, a reverse SPT rooted at the RP is constructed for the group as highlighted in Fig. 2.3.
- When the source wants to transmit data to a multicast group, the router associated with the source sends a *register message*, in which the multicast data is encapsulated, to the RP of the group. The RP receives the register message,

extracts the multicast data, and delivers the data downstream along the reverse SPT if corresponding forwarding states for the group exist. The RP may wish to send a joining message toward the source and establish forwarding states between the source and the RP, after which the RP can send a *register-stop message* to the source and receive the source's traffic without encapsulation.

Discussion. The DVMRP, PIM-DM, and MOSPF are also categorized as *dense mode* (DM) protocols, which are designed to perform best when the topology is densely populated with group members. The CBT and PIM-SM are normally categorized as *sparse mode* (SM) protocols, which are designed to work more efficiently when there are only a few widely distributed group members. The key disadvantage of the dense mode protocols is that states and routing computing are required in all routers in the network [4], regardless whether the router is on the multicasting tree. The sparse mode protocols typically offer better scalability where routing messages and forwarding states are only incurred at routers along the path between a source and a group member. Nevertheless, the core or RP could be a single point of failure or performance bottleneck; moreover, in the perspective of a source and receivers, having packets sent from the source to the RP and then to group members may cause nonoptimal routes to be adopted. Although PIM-SM provides a switch-over mechanism to switch from a shared tree to a SPT when a traffic rate threshold is violated, the operation complexity and messaging overhead are nontrivial [36].

2.1.2 Inter-Domain Multicast

The domain-based hierarchical architecture is the critical factor for the success of Internet [67, 74], which makes the routing scalable by route aggregation and offers policy-based routing to facilitate management. Thus, it is natural idea to extend the above-mentioned protocols, originally designed for a flat topology, into inter-domain scenarios. This thesis mainly focuses on the multicast in the inter-domain routing scenario, as the scalability is one of the most important concerns.

MBGP/PIM-SM/MSDP Approach. The multiprotocol extensions to border gateway protocol 4 (MBGP) has been developed to exchange network reachability information under multicast-specific policies. A particular solution is to extend PIM-SM for inter-domain multicast, where the multicast source discovery protocol (MSDP) is developed to detect active sources in each domain and announce the existence of the sources to all the RPs in different domains. The basic principles of MBGP/PIM-SM/MSDP approach are illustrated in Fig. 2.4:

- When a new source for a group becomes active, it will register with the domain's RP. In Fig. 2.4, H_1 registers to R_2 which acts as the RP in the domain.
- The MSDP peer (i.e., RP of the domain) can discover the existence of the new source and send a *source active (SA)* message to all directly connected MSDP peers. MSDP peers that receive a SA message will use a RPF controlled flooding

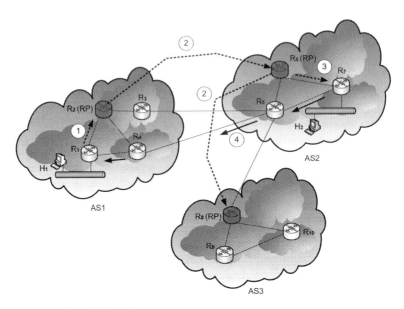

Fig. 2.4 MBGP/PIM-SM/MSDP

to have all RPs get the same SA message. In Fig. 2.4, R_2, R_6, and R_8 are MSDP peers, and the SA message from R_2 will be received by other two RPs.

- The RP that receives a SA message checks if it has any group members within the domain. If positive, the RP will extract the multicast data and transmit the data downstream the reverse SPT to the receivers. In Fig. 2.4, H_2 is one of the group members, and the R_6 delivers the multicast data packet to H_2's first hop router R_7.
- The designated router for a multicast member, for example, R_7, receives a multicast packet from H_1, and it sends a joining message on the RPF interface for H_1, which initiates setting up a source-based tree from H_1 to H_2.
- The role of MBGP is to provide control over the routing paths between domains. Suppose that in Fig. 2.4, the multicast traffic between autonomous system 1 (AS1) and AS2 should traverse the link between R_4 and R_5, instead of the link between R_3 and R_5. Since the joining message from the receiver always follows the reverse shortest path, there is a need to advertise a reverse path that is different from the unicast routing path selected by the joining message. The task is undertaken by the MBGP. With MBGP, R_4 can advertise itself as the next-hop for the reverse path from AS2 to AS1.

BGMP/MASC Approach. The border gateway multicast protocol (BGMP) [71] is a more general inter-domain approach. BGMP can construct a domain-level core-based bidirectional shared tree to connect multiple intra-domain multicast trees, with independent domain-specific multicast protocol allowable. To achieve this, the multicast address-set claim (MASC) [71] protocol is proposed to support the strict address allocation, where each domain is assigned a non-collision multicast address range. The address ranges are advertised throughout the Internet through BGP. To join in a group, the member of the group sends out the joining message toward the domain whose allocated address range includes the group's address. The joining message sets up forwarding states at BGMP-speaking border routers of each domain it passed. After the joining message arrives at the core domain, a PIM-like bootstrap protocol could be used to select the exact core. Intra-domain multicast (e.g., PIM and DVMRP) is used to forward data between the domain's group members and its BGMP border routers. These border routers in turn forward the data along the group's shared tree to reach members in other domains according to the states installed by joining messages. In contrast to *unidirectional* trees established by PIM-SM, where data can go only in a specified direction on each branch of the tree, data can flow in either direction along the bidirectional tree. For example, the data packet can be delivered to some receivers before it reaches the core.

The configuration complexity and poor scalability hinder either MBGP/PIM-SM/MSDP or BGMP/MASC approach to be a widely accepted inter-domain multicast mechanism. In the MBGP/PIM-SM/MSDP approach, information about the existence of all sources must be flooded throughout the Internet before multicast forwarding states can be installed. This extra complexity not only increases the overhead of managing groups especially when groups are dynamic [4] but also leaves little opportunity for scalability considering the bandwidth consumption in flooding. Moreover, the configuration complexity of MBGP is also nontrivial [31]. In the BGMP/MASC approach, the pillar scheme of strict address allocation itself presents a significant challenge. It is difficult to estimate how many multicast addresses are required in each domain; in addition, the dynamic joining and leaving operations could result in the frequent reallocation of the address space, and each reallocation result has to be propagated all over the Internet through BGP.

In the proposed AOM, regular BGP routing information is incorporated in the RPF concept to address the asymmetric inter-domain routing issue, which avoids the cost of deploying/configuring the MBGP and enables a fast group joining scheme. AOM takes a different perspective to investigate how to avoid the group-specific routing processing and forwarding states by leveraging the unicast routing.

Source-Based Model. With an argument that that many large-scale applications only require delivery from a single, often well-known source, *source-based* service model has been proposed [9, 37], where each receiver sends joint message directly to the source node to construct the tree. Source-based model does not induce the issue of RP discovery and significantly simplify inter-domain multicast. The source-based tree can also support many-to-many multicast by selecting one source as the route, constructing a bidirectional tree, and attaching other source nodes to the tree [4, 65]. A representative example is PIM-SSM (source specific multicast) [9], where the traditional many-to-many service paradigm is simplified as one-to-many paradigm. In PIM-SSM, the inter-domain tree for forwarding multicast data packets is rooted at the source, and the tree is constructed using the PIM-SM protocol. PIM-SSM is specifically suitable for subscriber-based systems that use logical channels. A *channel* is identified by a *(S, G)* pair, where *S* is a source

address and G is an SSM destination address, which is within the 224/4 range [9].
A SSM receiver application must know both the S and G before subscribing to the
channel. A number of techniques, including via Web pages, sessions announcement
applications, etc. are available to let receivers know the channel information. The
receiver then subscribes to the channel by explicitly initiate a (S, G) join, instead of
a $(*, G)$ join in the many-to-many model.

The proposed AOM also adopts the source-based service model; however,
it incorporates the feature of *localized group identifier (ID) allocation* and
decouples the group ID with multicast routing. Thus, AOM can further enhance its
scalability and management flexibility in supporting a large number of multimedia
communication groups.

2.2   Overlay Multicast

IP multicast has not been widely deployed due to its scalability issue and other
technical and marketing reasons [54, 81], which inspires researcher to implement
multicast at the end systems, that is, overlay multicast. Many overlay multicast
mechanisms have been advocated in the past years. The primary approaches
for overlay multicast are categorized by how the data delivery structures are
constructed.

2.2.1   Approaches for Overlay Multicast

Construct Trees Through Meshes. Narada [81] and Gossamer [16] construct trees
in two logical steps, where a strongly connected mesh structure is first constructed
through the bootstrap scheme, then a spanning tree structure is built based on the
mesh. Figure 2.5 illustrates this process. When a member wants to join a group,
it acquires a list of group members from a special server or member discovery
mechanism. The newcomer selects a subset of the existing members as its mesh
neighbors as shown in Fig. 2.5. Members in the mesh exchange refresh messages
periodically to maintain or optimize the mesh. On top of the mesh, the group runs
an application-level routing protocol similar to DVMRP to compute source-based
reverse SPTs, as highlighted in Fig. 2.5. A significant difference between Narada
and Gossamer is that the nodes building the mesh in Narada are end hosts, while in
Gossamer are application servers deployed in the network infrastructure.

Construct Trees Directly. Some overlay multicast schemes directly construct data
delivery trees [33, 40, 41], where each group member is responsible for finding its
appropriate parent on the shared tree. A joining member first queries the RP server,
which responds the list of members already in the group [33] or the root of the
shared tree [40, 41]. In the first case, the joining node will probe all the members
in the list and choose the one that is best in the metric of interest as its parent.

Fig. 2.5 Construct the tree
through mesh

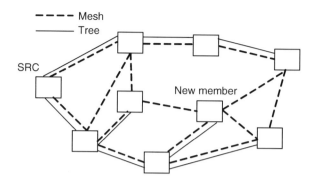

In the second case, the joining member will examine the entire tree starting from the root to find a member that is closest (in terms of hops or round-trip time) to itself. The shared tree has a degree bound, which limits the number of children to be supported. When the joining member sends requests to the potential parent it chose, the latter will decide whether to grant the request according to its policy (e.g., bandwidth, traffic load, etc.). In order to measure the metric of interest, tools for investigating the network information such as *traceout* have to be available. While the above mechanisms compute the shared tree in a distributed manner, ALMI [75] proposes a completely centralized solution to construct the tree. All responsibilities for group management and overlay maintenance are taken by the central controller, which is a program instance and located at a place that is easily accessible by group members.

Structured Overlay Facilitated Approach. Hypercast [52], Delaunay Triangulations [60], CAN [47], Bayuex [42], and Scribe [48] build overlay networks by assigning to members logical addresses from some abstract coordinate space. The neighbor relationships are regulated by addresses assigned to members. With these logical addresses, members are organized into various topologies [46]. For example, Hypercast assigns each member of the overlay network a binary string and builds an overlay network with a hypercube topology. As the name suggests, Delaunay Triangulation protocol constructs the overlay network as Delaunay Triangulations, where for each circumscribing circle of a triangle formed by three node in the node set, no node in the node set is in the interior of the circle. CAN-based multicast assigns logical addresses from Cartesian coordinates on an *n-dimensional* torus. The next-hop routing information for multicasting can be encoded in the logical addresses thus, the routing protocol for the overlay is not required.

Figure 2.6 illustrates the basic principles of CAN. The constituent nodes of the CAN form a virtual two-dimensional Cartesian coordinate space, where nodes A–E are labeled in their respective zones. When a new node Z wants to join the group, it first finds a node already in the CAN through some bootstrap mechanism. In Fig. 2.6, Z finds B, which provides Z a list of participating nodes. Z randomly chooses a point in the coordinate space, which is owned by E in the example. The joining request is then sent to E, then the area of E is split and half of the area belongs

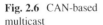
Fig. 2.6 CAN-based multicast

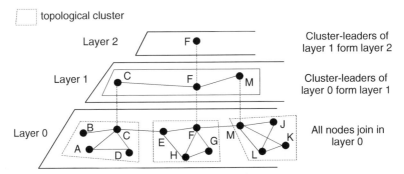

Fig. 2.7 Hierarchical arrangement of hosts in NICE

to the joining member Z. In order to realize multicast, the participating nodes just form group-specific mini-CANs using the similar procedure as in Fig. 2.6, where the constituent nodes of the mini-CAN are members of the group. The multicasting data delivery is achieved by efficient flooding over these mini-CANs [47].

Hierarchical Cluster-Based Approach. In order to achieve better scalability, Nice [8] and Kudos [78] organize members into hierarchies of clusters. The network architecture of Nice is shown in Fig. 2.7. All members join in layer 0, where the members are partitioned into a set of clusters based on the topology. The cluster leaders of layer 0 form layer 1, where the leader is the host with the minimum maximum distance to all other members in the cluster. Similarly, the cluster leaders from a lower layer form the layer above. When a new member wants to join the multicast group, it first queries a RP, which is a special host for protocol initialization. The RP responds with the hosts currently in the highest layer of the hierarchy. The joining member then contacts all members in the highest layer to find the member closest (in terms of end-to-end latency) to itself. Iteratively, the joining member keeps identifying the closest member to itself at each layer until it is mapped to some cluster in layer 0. The member hierarchy is used for multicast data delivery and maintained by exchanging periodic refreshing messages. Kudos uses the similar hierarchy, but it constructs only two-layer hierarchies with a Narada-like protocol at each layer.

2.2.2 An Architectural Perspective

An important aspect of overlay multicast proposals is the network architecture, that is, whether they emphasize a peer-to-peer architecture or an infrastructure-based architecture. In the overlay with a peer-to-peer architecture, data forwarding in overlay networks is performed at the pure application level, while in the infrastructure-facilitated architecture, the overlay networks consist of a collection of proxies or servers placed at strategic locations in the network fabric. For example, Yoid [33] and ALMI [75] emphasize the peer-to-peer architecture, but Gossamer [16] and Overcast [41] emphasize the infrastructure-facilitated architecture. In both cases, data packets may traverse the common transport network link multiple times before arriving at the destinations, which incurs inefficient use of network capacity and increased delays compared to IP multicast. In order to improve the efficiency, most overlay multicast protocols based on peer-to-peer architecture implement schemes to collect end-to-end measurements among members and build the overlay structure according to these measurements. In the infrastructure-facilitated architecture, an optimal overlay can be constructed with the embedded support in the IP routing infrastructure. An efficient overlay multicast cannot be accomplished with ignorance of the network layer intelligence.

Both the end-to-end measurements and infrastructure facilitation need application-specific processing from within the network. In this sense, the multicast issue by nature requires incorporating intelligence from both application and network layer. That is, identifying the users associated with a multicast group requires application-layer membership management, while delivering data to the proper destinations needs network-layer support according to the application-layer membership information. Our perspective is that multicast should be one of the infrastructural functionalities provided by next-generation Internet. Multicast is a generic service paradigm which could be commonly used by the ever-growing multimedia applications (e.g., online multiplayer games, IPTV, video conferencing, etc.), and incorporating the commonly used service by multiple applications is a primary requirement for next-generation Internet, which is conceived as a general-purpose, shared infrastructure [22]. Moreover, the improvements in computation power and cost-efficiency enable the networking devices to execute more complex software and assume more computation burden, which provides an opportunity to streamline the design of networking functionalities for next-generation Internet. The difference between the infrastructure-facilitated overlay multicast and our scheme is that the former maintains a clean separation between the packet routing infrastructure and the overlay facilitating entities, while the latter embeds the application-level intelligence into the network routers.

2.3 Modern Multicast: Exploiting the Enhanced Intelligence

In most of the modern multicast protocols, the group-specific forwarding states maintained at each router (as adopted by those legacy IP multicast protocols) are traded with packet-carried information and processing/computation at each router for better scalability. The common feature of these mechanism is that network routers are enhanced with more intelligence so that they can have access to more information in the packet and perform more complicated computation in addition to looking up the routing table and forwarding.

2.3.1 REUNITE

Recursive Unicast Approach to Multicast (REUNITE) [31, 62] stores forwarding states only at "branching nodes," where the multicast flow diverges. The basic principle of the REUNITE protocol is to use unicast to implement multicast service. In the joining process, receivers send explicit joining messages toward the source, but their addresses are to be maintained at different branching nodes in the format of a receivers list. The first subscriber's address will be maintained by the source. The forwarding states are appropriately set up so that each receiver's address is maintained at exactly one node and a source-based data delivery tree is established at end. To multicast a packet, the source sends the packet to the first-subscribed receiver. When a branching node forwards such a packet, it sends a copy of the packet to each receiver in its own receivers list. This procedure continues recursively until packets reach all receivers.

An example of REUNITE is given in Fig. 2.8, where a packet is multicast to three receivers. S is the source, the list of receivers maintained by each node is shown in the figure. Note that R_4 and R_6 are branching nodes, and forwarding states are only maintained at these two nodes in the network. When S multicasts a packet, it simply unicasts the packet to all receivers in its list (i.e., H_1). R_2 and R_3 just act as normal unicast routers. When R_4 forwards the packet, it also sends a copy to H_3, which is the only element in its list. Finally, when the packet goes through R_6, R_6 makes another copy and sends it to H_2.

The forwarding states reduction of REUNITE may be counteracted when the number of branching nodes grows as the size of the multicast group increases. In this case, the number of forwarding states also increases. Moreover, the branching node of the data delivery tree still maintains per group states, which leads to the scalability issue. Multicast forwarding tables in REUNITE are stored with soft states, which induces large messaging overhead to refresh the destination information maintained in the branching nodes. The proposed AOM stores destination-specific, instead of group-specific forwarding states in each router, where the forwarding complexity is totally independent of the number of groups to be supported, resulting in desirable scalability.

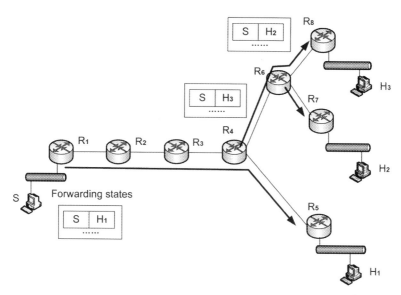

Fig. 2.8   REUNITE

2.3.2   Xcast

Explicit multicast (Xcast) [11] eliminates the need for routers to maintain multicast forwarding states by storing a list of all receivers' IP addresses in the multicast packet header. When a Xcast router receives a packet of this kind, the router will check its local unicast routing table to determine the next-hop for each destination in the list. Destinations with a common next-hop are aggregated as a subset. For each distinct next-hop, the router generates a packet replica and modifies the list of destinations in the packet replica so that the list only includes the destinations that would be dispatched through this next-hop. The packets replicas are then delivered along those next-hop interfaces. The Xcast uses the tuple *(source address, channel identifier)* to define a channel; therefore, the global multicast address allocation can be avoided.

The Xcast is scalable in terms of the number of groups that can be supported since the routers in the network do not need to store and maintain any multicast routing information for these groups; however, this advantage comes at a cost that the per packet processing complexity in each router is increased. Moreover, the Xcast is incapable of supporting an application where there is a large number of subscribers for each group. This is because the maximum number of receivers that can be carried in one packet is restricted by the limited length of the Xcast header. The proposed AOM takes the same service model as Xcast; however, AOM utilizes bloom filter technique to constrain the overhead incurred by the explicit addressing, thus significantly improves the scalability and bandwidth efficiency.

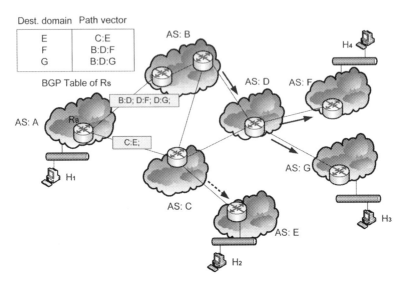

Fig. 2.9 Free riding multicast (FRM)

2.3.3   FRM Family

Free riding multicast (FRM) [30] needs no forwarding state at routers either, which is realized by encoding all source-based tree branches for a group into a bloom filter [13] and carrying the bloom filter in the packet to facilitate multicast forwarding. With FRM, the border router of each destination domain will piggyback the active group information, encoded in a bloom filter [13, 17], on its regular BGP advertisements to reach the source domain. Based on the information, the border router of the source domain can detect the destination domains for an active group and compute a domain-level multicast tree for that group. The multicast tree will then be encoded using another bloom filter (shim header) and inserted into each multicast packet. When it receives a multicast packet, each router will examine its BGP forwarding entry against the attached bloom filter to determine the output interfaces.

Figure 2.9 illustrates the forwarding process of FRM. A packet multicast to a group by host H_1 is first sent to the border router of the domain R_s. From the BGP advertisements, R_s knows that domains E, F, and G have members in the group and computes the multicast tree from its local BGP table from domain A to each of the domains above. R_s then forwards a packet copy to domain B along with an encoded subtree highlighted by arrows and another copy to C with an encoded subtree in dotted arrow style.

The performance of FRM depends on the shape of the multicast tree. It is suitable for the tree with rare branches as the branches have to be carried by the packet. If the number of branches exceeds the capacity of the packet header, multiple packets will be sent out to cover all the subscribers in the group, which incurs redundant traffic. Another drawback is that FRM can induce long joining delay because receivers have to subscribe to a group through the processing at the border router of the source domain, even if the joining message may have passed some domain that is already on the multicast tree. Moreover, the border router of the source domain may become a performance bottleneck as all the tree computation burden is undertaken by the router.

A critical issue of FRM is that the false-positive inherent in the Bloom filter may incur forwarding loops, which is taken into consideration by its later variants. LIPSIN proposes to deploy a FRM-like multicast protocol in a publisher/subscriber network fabric [68]. Each link of the network is assigned d IDs. Thus, there are d candidate in-packet Bloom filters for a given multicast tree, from which a Bloom filter with the best false-positive performance can be selected. When receiving a packet, the router analyzes the in-packet Bloom filter to check if it contains a path that may lead the packet to return. If positive, the packet and its incoming interface will be cached. A loop is detected if the packet with the cached in-packet Bloom filter returns to the router from an interface other than the cached one. Nevertheless, the router caching the suspect packet is not necessarily the origin of the loop; therefore, the false-positive traffic cannot be fully truncated. To deal with the challenge, the caching router has to signal a request upstream toward the data source to insert a different Bloom filter in the packet for the multicast tree, which imposes much burden on the data source node; moreover, there is no way to guarantee that the reencoded Bloom filter will never incur forwarding loop at a different router in the network.

BloomCast proposes a *bit permutation* technique to reduce the Bloom filter false-positive effect. Different from FRM's directly encoding tree branches at source node, BloomCast let joining messages record each hop they traveled starting from leaves of the tree, encode the hop in a Bloom filter, and remap the Bloom filter to a different arrangement at each intermediate router. A unique reverse SPT is then created at the data source node by ORing all cumulatively permuted Bloom filters in joining messages. During the forwarding, the falsely delivered packet cannot be correctly de-mapped through the bit permutation at each hop, so the packet with no matched output interfaces will be dropped. Unfortunately, although BloomCast works smoothly under the symmetric routing assumption, where the shortest path from node A to B is the same one used to go from B to A, the inter-domain routing is usually asymmetric for the administrative reasons [31]. Moreover, BloomCast still cannot identify the origin of the forwarding loop once it occurs, and bit permutation can only mitigate the probability of the forwarding loop rather than totally prevent it.

The proposed AOM also uses the bloom filter technique, while it takes a protocol independent approach to build up the routing information. Basically, the membership report message travels to the source against the unicast routing table and serves to control the operations of AOM (e.g., joining, leaving). The membership message will also install forwarding states along its path in a compressed manner by a Bloom filter. In contrast to FRM that carries the domain-level multicasting tree in the packet, AOM encodes the destination network prefixes into a Bloom filter and attaches the bloom filter to each multicast packet by the source node for RPF. In addition to the advantages of lighter bandwidth overhead and shorter joining delay, compared to FRM, AOM does not need the border router of a source domain to do any group-specific computing, which can significantly improve the scalability in the case that a source domain may include a considerable number of service providers (SPs), each of which further offers thousands of channels. Moreover, AOM can automatically eliminate the forwarding loop caused by the Bloom filter false-positive without any assistance from the underlying infrastructure, compared with LIPSIN and BloomCast.

2.3.4 MAD

A novel architecture named multicast with adaptive dual-state (MAD) [66] is to limit the number of routers that have to keep multicasting group states. In MAD, for a given group, routers are firstly mapped as an overlay core-based tree named as the membership base tree. When a subscriber wants to join in a group, it sends a joining message toward the core along the base tree until it reaches the first node that, already on the membership tree (MT). An *en route* overlay router joins the membership tree (MT) whenever the subtree rooted at this node in the base tree has at least a minimum number of firs-hop routers with attached subscribers. Each node on the MT keeps the states of *mtChild* (i.e., children of this node in the base tree that belong to the MT) and *mtFHs* (i.e., a list of first-hop routers with attached subscribers that are downstream of this node in the base tree). When a source wishes to multicast a packet to a group, it first forwards the message to the core of the group in the MT. When the core receives the message, it forwards the message to all the logical routers recorded in the *mtChild* and *mtFHs* through overlay multicast. Each router receives this packet conducts the same operation until all subscribers receive the packet.

For example, consider the topology in Fig. 2.10a, the traditional IP multicast tree requires 11 routers to maintain the multicast membership states, while the MAD MT only needs 4 routers to do the job as shown in Fig. 2.10b. In the example, when the first subscriber H_1 comes, its joining message is delivered to the core A and A adds M, H_1's first-hop router, to its *mtFHs*. When the subsequent receiver H_2 joins, A discovers that the subtree rooted at C in the base tree has two first-hop routers, which exceeds the threshold in this example. A informs C to create membership states, which will include M and O into its *mtFHs*. Meanwhile, A records C in its *mtChild*. Similarly, when H_3 and H_4 come, B and E create membership states, where B will include J and K in its *mtFHs*, and B will set E in its *mtChild*. The joining of H_5 will add an entry I in the *mtFHs* of B. When multicasting, A dispatches the packet copies to B and E. When B receives the packet, it further generates packet copies and delivers them to E and I, which are in its *mtChild* and *mtFHs*, respectively.

The key idea of MAD is the decoupling of group membership and forwarding state, which is achieved by leveraging the overlay MT. Instead of storing *(Group*

Fig. 2.10 MAD trees

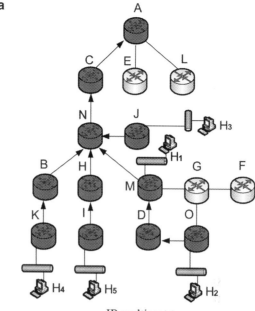

IP multicast tree

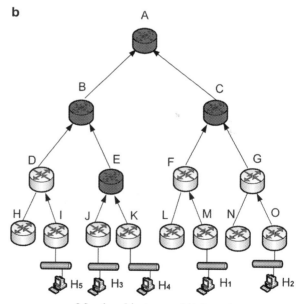

Membership tree and its base tree

ID, output interface list) as in traditional IP multicast, MAD keeps the information of the routers that are significantly involved in the multicast in the perspective of MT. However, establishing a MT for every group introduces extra complexity and messaging overhead. The reduction of membership states is at the cost of bandwidth consumption because joining messages and multicast forwarding are implemented using overlay multicast, which could incur redundant traffic at the network layer. To achieve different optimization objectives, MAD needs to adaptively switch between overlay multicast and IP multicast mode, depending on the service scenarios. Such a scheme can mitigate memory overhead when there are a large number of groups with infrequent data traffic but cannot fundamentally solve the scalability issue. In AOM, the membership management component and the multicast forwarding component are also completely decoupled; moreover, AOM provides a uniformed solution for all application scenarios.

2.4   Application-Oriented Networking

This section first explain the AON concept in detail and then introduce the possible technique enabling AON implementation.

2.4.1   AON Concept[1]

The retrospect of multicast evolution reveals a technical trend that the design of multicast trades in the efficient link bandwidth and simple per packet processing in the network for less states storage and messaging overhead in routers. While the REUNITE allows the router in the network to initiate multiple unicast to subscribers, the Xcast enhances the network router with the capability of replicating packets and modifying the contents in them. While the FRM encodes the multicasting tree into the packet and tilts more computational complexity to routers, the MAD organizes the network routers into an overlay tree structure. It is notable that more and more intelligence supposed to be provided in the application layer is incorporated in the network layer, where network routers can do more than looking up the destination address in the unicast routing table and forwarding.

Integration of application intelligence into the network or allow application-specific computation in the network is not a brand new idea, which has been taken as an efficient approach, implicitly or explicitly, to implement some basic networking functionalities develop new network protocols, or facilitate upper-layer applications. For example, the domain name service (DNS) and dynamic host configuration protocol (DHCP) are fundamental networking functionalities, but implemented through application-layer processing. In recent decades, along with the global popularity of Internet and the proliferation of various multimedia and security applications, we have seen more and more application-specific nodes,

---

[1] ©[2008] IEEE. Portions reprinted, with permission, from [72].

e.g., Web proxies, multimedia gateways, and wireless access gates, being inserted into the network [55, 57, 58]. The IP multicast is an example of integrating application intelligence into routers since it has been proposed [4, 49]. However, all of these application-oriented solutions had been implemented in an ad hoc manner [34, 45, 45, 56].

Motivated by the necessity of allowing network to perform customized computations, the *active network* [70] was proposed in the mid-1990s as a generic architecture to provision programmability within the network, instead of those ad hoc approaches. In an active network, packets are replaced with *capsules*, which are program segment (possibly with embedded data) executable by an active network node. The active network has never been widely deployed; the main reasons include the large bandwidth overhead of carrying programs, lack of a common capsule program language, and the security issue due to users' active control capability.

With the Internet evolving into an extremely complex system, developing new Internet applications by coding from scratch becomes inefficient and impractical. The service-oriented architecture (SOA) [64] is being widely adopted as a scalable approach for service creation according to a "find, bind, and execute" paradigm. In SOA, complex services or applications can be assembled on demand by combining necessary service components. Such SOA-based service creation is implemented by exchanging messages, described in eXtensible Markup Language (XML), among service requesters, registries, and providers.

The efficient implementation of SOA requires a powerful messaging backbone, which is one of the major motivations leading to the Cisco AON technology [19]. An AON-based network can transparently intercept the content and context of application messages and conduct operations on those messages according to business-driven policies and rules. The AON concept is closely related to the active network architecture; an XML message can be interpreted as a capsule capable of activating in-network processing. However, with AON, it is the network itself that determines what application-level processing capabilities to be offered within the network. By limiting users' control (available in the active network), the network-controlled approach not only facilitates the deployment of application-oriented capability in a coordinated manner but also considerably improves the security level of the whole system.

The current AON studies focus on upper-layer applications [20, 21]. How to systematically exploit the AON capability to enhance the network functionalities is currently obscure in both industry and academia. In this research, we initiate the study in this area by proposing an AON-based multicast scheme, which is formally termed as application-oriented multicast (AOM).

Before discussing the details of the multicast scheme, we first present a streamlined architecture of an AON router, as illustrated in Fig. 2.11. In an AON router, the incoming traffic will be first classified as *normal traffic*, which does not need application-level processing and is directly forwarded against the IP routing table, and *AON traffic*, which requires application-level processing before forwarded. The AON traffic will be further categorized and dispatched to different application-specific AON modules. We can select 1 bit in the IP header, for example, one of the

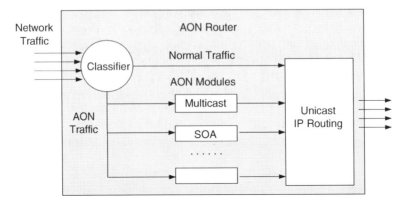

Fig. 2.11 A generic architecture for an AON router

type of service (TOS) bits in the IPv4 header or one of the Traffic Class (TC) bits in the IPv6 header, to behave as the normal/AON traffic indicator flag. The flag is set to "1" for indicating the AON traffic. Although more TOS bits and TC bits may be used to further identify the AON modules, we prefer that the fine-grained classification information is carried in the payload for higher scalability and flexibility.

2.4.2 Enabling Technique

In order to implement AON concept, the router has to access more information in the data packet and perform more complicated processing. The deep packet inspection (DPI) technique can be used to enable the required functionalities.

DPI refers to technologies that inspect not only the packet header but also the packet payload, which enables many new network services that motivate AON [28]. For example, enterprises with DPI could realize intelligent packet switching and routing, strong security system, effective traffic management, and quality of service (QoS) assurance; internet service providers (ISPs) could apply DPI to facilitate content-based billing, differentiated services, and enhanced traffic monitoring. In brief, the DPI technology strengthens the visibility, control, and service creation capacities of the global IP networks.

DPI has to recognize numerous Internet and network applications and protocols to provide application awareness. Applications and protocols have their distinguishable behavior patterns and properties, which can be used to identify themselves. The specific pattern and property that can uniquely identify an application or protocol is referred to as the signature of the application or protocol. There are three main methods to classify the signatures, based on how deep the packet is examined [29].

Analysis by Numerical Properties. An application or protocol will present some special numerical properties during its operation. Investigating the numerical

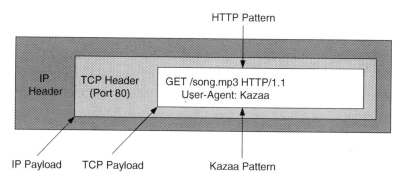

Fig. 2.12 Kazaa string match analysis

characteristics within a packet or several packets could help classifying the traffic. These properties include the payload length, the short-term packet length histogram, and the number of packets sent in response to a specific interaction. For example, Skype (versions prior to 2.0) establishes connections with UDP messages. The client first sends an 18-byte message, and an 11-byte message will be returned from the server. The client then sends a 23-byte message, and the server sends back a message of 18, 51, or 53 bytes in response. If such a behavior pattern appears, it is probably that the traffic is from a Skype application. Another example, pure HTTP packets and P2P-related traffic have different packet length histograms: pure HTTP packets concentrate around the size of several hundred bytes, while P2P-related control packets are shorter, normally about a hundred bytes long. Numerical property analysis does not touch the deep side of the packet, the signature identification based on which may not be accurate enough.

Analysis by Port Number. Port analysis is based on the fact that an Internet application is normally associated with one or multiple port numbers. For example, an incoming e-mail (POP3) typically uses the port 110 (995 if it is secure), while the outgoing e-mail (SMTP) uses port 25. Port analysis is an easy and well-known method for signature analysis; however, it is a weak method at the same time. This is because some applications use random ports instead of fixed/default ones, and an application can arbitrarily select its associated port number. For example, Skype often uses port 80, but it also uses port 443, which is supposed to be used by secure Web applications. The Skype even allows users to choose certain port numbers themselves.

Analysis by String Match. Seen from the depth aspect, while the numerical property analysis inspects exterior behaviors of the packets, and the port analysis examines a specific section of the payload at the transport layer, the string match analysis goes deeper to the inspect the payload at the application layer to search for textual characters or numeric values. These strings may reside in a fixed position or distribute at different locations within the packet. For example, Kazaa presents the protocol name in the packet payload using a typical HTTP GET request, as illustrated in Fig. 2.12.

String match is important for more accurate signature classification. In Fig. 2.12, if analysis is performed by port analysis alone, the port 80 and HTTP GET request may indicate the packet as from Web-browsing HTTP traffic, but in fact, it is generated by the Kazaa application.

Many industrial DPI players and academia researchers are active in the community. There are many commercial products and related research results that can be utilized to implement AON concept with reasonable cost.

Bivio Networks' 2000 and 7000 series use DPI for application acceleration while attempting to keep their costs comparable to typical appliances [10]. Cavium Networks [15] and LSI [53] Tarari make specialized processors that are bundled with other vendors' products, such as firewalls and filtering devices. Allot [3] NetEnforcer Series provide granular visibility and dynamic control over wide area network (WAN) and broadband services based on DPI technique. Bivio Networks, Allot Communications, and LSI Corporation, along with some other companies, founded dPacket.org to foster and support community interest and progress in DPI. dPacket.org's primary mission is to serve as a centralized platform for both the networking community and broader stakeholders to inform themselves and collaborate on addressing challenges from technical, operational, ethical, and legal perspectives.

In academia, most of the efforts are put to develop efficient string matching algorithms to check the content of the packets against a set of predefined strings so that some anticipated patterns appearing in the packets can be recognized [29]. Many of these algorithms are designed for the purpose of intrusion detection [1]. These algorithms can be categorized according to the implementation as software-based, hardware-based, or mixture of the two. The most well-known software-based algorithms are Knuth–Morris–Pratt (KMP) [50], Boyer–Moore (BM) [12], Aho–Corasick (AC) [2], AC–BM algorithm [23], Wu-Manber [79], and Commentz Walter (CW) [24]. The popular hardware-based algorithms include parallel Bloom filters [27], content addressable memory (CAM), ternary CAM (TCAM), and field-programmable gate array (FPGA) implementations.

2.5 IPTV Over Multicast

IPTV delivers TV streams as a series of IP packets using multicast. Chapter 5 of the book will study the application of the proposed multicast mechanism to IPTV service; the background of the IPTV system is briefly reviewed in this section so that the description in Chap. 5 can be understood.

The general definition of IPTV by the International Telecommunication Standardization Sector focus group on IPTV (ITU-T FG IPTV) is [18]: *An IPTV service (or technology) is the new convergence service (or technology) of the telecommunication and broadcasting through QoS controlled Broadband Convergence IP Network including wire and wireless for the managed, controlled and secured delivery of a considerable number of multimedia contents such as Video, Audio, data*

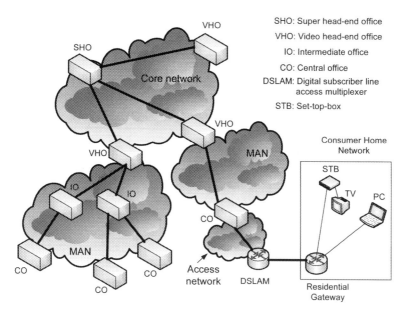

Fig. 2.13 A typical IPTV network architecture

and applications processed by platform to a customer via Television, PDA, Cellular, and Mobile TV terminal with STB module or similar device. This definition indicates that the transport network of IPTV should be IP-based and guarantee the QoS and QoE. The network of IPTV is managed and operated by the IPTV carrier and cannot be accessed without authorization. The IPTV service is different from some popular peer-to-peer applications over the Internet such as PPLive and PPstream, which are normally categorized as *Internet TV* [38]. The Internet TV is open to any user, but the QoS and QoE are not guaranteed.

2.5.1 IPTV Architecture

A typical IPTV network architecture is illustrated in Fig. 2.13 [69, 77]. The super head-end office (SHO) is the primary source of nationwide content, which could be from the service provider or satellite. The content will be digitally encoded or reencoded into flows of IP packets. Those flows are transmitted to video head-end office (VHO) through the high-speed core network. Each VHO is responsible for subscribers within a metropolitan area network (MAN). VHO can encode/reencode the local video content (e.g., local news) and perform some commercial processing (e.g., advertisement insertion). Then the nationwide and local TV programs are transmitted to subscribers upon request over the MAN. The TV programs may need to go through intermediate office (IO) and central office (CO) before reaching the

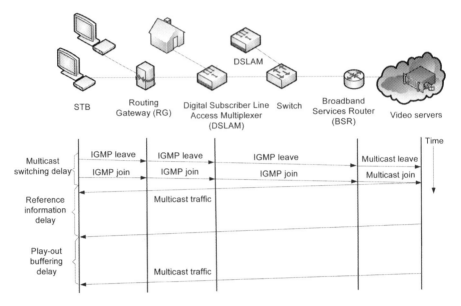

Fig. 2.14 IPTV channel zapping

end users' access network. IPTV carrier will decide if IO and CO are necessary. After going through the access network, the TV program arrives at the digital subscriber line access multiplexer (DSLAM). Eventually, the end users can watch the program with the assistance of consumer home network. In order to save bandwidth consumption in the network, those IP flows of TV programs are delivered from SHO to consumer home network via IP multicast.

2.5.2   IPTV Channel Zapping[2]

The major process of channel zapping in a typical IPTV access network architecture is illustrated in Fig. 2.14 [6,39,43,44,51,76]. IPTV system uses multicast to deliver data when a user requests to switch to a new channel, the STB requests to leave the current channel and join in the new one through the IGMP signaling [14, 32, 80]. The broadband service router (BSR) is the IGMP router, which will establish/prune the tree branches for multicast groups according to the IGMP information.

   In practice, when the multicast stream reaches the STB, the streaming data is encoded in some container format and cannot be directly decoded by the decoder [6]. The receiver has to acquire and parse certain reference information before it can process the multicast traffic. The reference information includes control

[2]©[2012] IEEE. Portions reprinted, with permission, from [73].

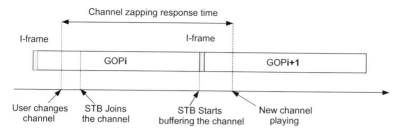

Fig. 2.15 IPTV channel zapping response time

information and the I-frame. The control information must be first obtained, with which the receiver knows how to decode the data from the container format and find the audio/video elements or security-related information for the multicast stream [6, 35, 76]. With the obtained element video stream, the receiver has to wait until the appearance of the first I-frame before the decoder can recover pictures from the data stream. The reference acquisition process must be finished first, and then the generated decoding-friendly streaming data can be available for play-out buffering, where the network jitter is smoothed out and some application-specific functionalities are implemented [6, 76].

The time interval from the IGMP leave message sent out to the STB joining in the new channel is typically short and not the primary factor of the channel zapping time [6, 7, 39, 76]. The buffering delay is decoder-implementation specific and unavoidable for decoders. In the reference information delay, the time for acquiring the I-frame is a major contributor [7, 76, 77]. The IPTV video object is normally encoded as series of GOPs [77], with each GOP starts with an I-frame. An I-frame contains the representation of an entire picture, and the receiver can recover a GOP only after the corresponding I-frame is acquired. GOP durations can be up to a few seconds, and the joining request for a zapping act can arrive right after the transmission of the starting I-frame; thus, the time between the arrival of the joining request and the next I-frame (i.e., FID) can be up to a few seconds, as illustrated in Fig. 2.15. FID isthe dominant channel zapping time contributor solvable through networking techniques [7, 76, 77].

Recently, IPTV zapping acceleration scheme based on time-shifted subchannels (TSS) is introduced [7]. In TSS-based model, an IPTV channel is companied with several subchannels that are spaced by T time units, and each subchannel is a full-quality replica of the main channel media stream. The subscriber trying to change channel first selects a subchannel that offers the earliest I-frame and then migrates to the main channel when enough data packets are accumulated. The TSS-based model is more advantageous than most of existing channel acceleration schemes based on the auxiliary stream; however, the TSS-based model implemented with IP multicast will incur frequent messaging overhead and will be easily affected by the joining message signaling time in practice.

The fundamental reason is that the multicast identifier is associated with the data forwarding in IP multicast. The receivers have to locally decide the subchannel to join in, and the data server has to broadcast to all receivers specifying which subchannel is providing the earliest I-frame. Moreover, the channel zapping performance is easily affected by the joining message signaling time under IP multicast, which may cause the subscriber to miss the expected I-frame over the selected subchannel. In addition, with IP multicast, the subscriber migration act is performed with several rounds of join-and-leave operation, which incurs high risk of system errors. Further, the coupling of multicast address and data forwarding leads to inflexible and complicated group ID management in IPTV system.

2.6 Summary

Since the multicast issue came up, leveraging the unicast routing information has been the cornerstone idea underpinning various multicast protocols. There have been three basic approaches to exploit the unicast routing for multicast. First, RPF against unicast table can construct a source-rooted SPT in a *broadcast-and-prune* manner. Second, MOSPF exploits OSPF to flood an OSPF area with the group membership information so that each MOSPF router can independently construct the shortest-path tree for each source and group. Third, each receiver can send explicit joining message to a selected core or rendezvous point (RP), along the unicast shortest path between them, to construct a bidirectional shared tree. The domain-based hierarchical architecture is the critical factor for the success of Internet, which inspires researchers to extend the above-mentioned protocols, originally designed for a flat topology, into inter-domain scenarios, for which the approaches of MBGP/PIM-SM/MSDP and BGMP/MASC are proposed. However, the complexity of these two approaches leads to a *source-based* service model which simplifies the original many-to-many multicast communication paradigm into a one-to-many model.

Limited deployment of IP multicast stimulates the development of overlay multicast. Existing overlay multicast schemes can be categorized based on how the overlay structure is constructed. First, the overlay multicast trees can be constructed through meshes. Second, trees can also be constructed directly. Third, structure overlay approach. Fourth, organize group members to form hierarchical clusters. Efficient overlay multicast needs to combine both application and network layer intelligence.

The development of modern multicast mechanisms reveal a technical trend that the design of multicast trades in the efficient link bandwidth and simple per packet processing in the network for less states storage and messaging overhead in routers. This is achieved primarily by enhancing the capability of network routers with application-layer intelligence, which is the common characteristic of more recently proposed multicast schemes including REUNITE, Xcast, FRM, and MAD.

Integration of application intelligence into the network or allowing application-specific computation in the network has been taken as an efficient approach, implicitly or explicitly, to implement some basic networking functionalities, this approach is termed as AON technique. This research initiates the study on how to systematically exploit the AON technique to improve the network functionalities by focusing the multicast.

An important application of multicast is to facilitate IPTV delivery. IPTV encodes TV program videos into a series of segments and delivered these segments via IP multicast. The typical network architecture of IPTV has been introduced in this chapter. The IPTV in nature has the channel zapping issue, where the act of changing a channel to another may take a long time. The channel zapping issue has been explained in detail.

References

1. AbuHmed T, Mohaisen A, Nyang D (2007) A survey on deep packet inspection for intrusion detection systems. Mag Korea Telecommun Soc 24:25–36
2. Aho AV, Corasick MJ (1975) Efficient string matching: an aid to bibliographic search. Commun ACM 18:333–340
3. Allot, http://www.allot.com/. Accessed Jan 2009
4. Almeroth KC (2000) The evolution of multicast: from the mbone to interdomain multicast to internet2 deployment. IEEE Network 14:10–20
5. Ballardie A (1997) Core based trees (CBT) multicast routing architecture. IETF RFC 2201. http://tools.ietf.org/html/rfc2201
6. Begen A, Friedrich E (2009) RTP payload format for MPEG2-TS preamble. Internet-Draft draft-begen-avt-rtp-mpeg2ts-preamble-04, Oct 2009
7. Bejerano Y, Koppol PV (2009) Improving zapping response time for IPTV. Proc. IEEE INFOCOM, Mar 2009, pp 1971–1979
8. Bhattacharjee B, Banerjee S, Parthasarathy S (2001) A protocol for scalable application layer mutlicast. Tech. report, University of Maryland, Jul 2001
9. Bhattacharyya S (2003) An overview of source-specific multicast (ssm). IETF RFC 3569. http://www.ietf.org/rfc/rfc3569.txt
10. Bivio, http://www.isi.edu/newarch/iDOCS/final.finalreport.pdf. Accessed Jan 2009
11. Boivie R, Feldman N (2001) Explicit multicast (Xcast) basic specification. Internet Draft. http://tools.ietf.org/html/draft-ooms-xcast-basic-spec-01
12. Boyer RS, Moore JS (1997) A fast string searching algorithm. Commun ACM 20:762–772
13. Broder A, Mitzenmacher M (2004) Network applications of bloom filters: a survey. Internet Math 1:485–509
14. Cain B, Deering S, Kouvelas I, Fenner B, Thyagarajan A (2002) Internet group management protocol, version 3. IETF RFC 3376, Oct 2002. http://www.ietf.org/rfc/rfc3376.txt
15. Cavium, http://www.caviumnetworks.com/Table.html#NSP. Accessed Jan 2009
16. Chawathe Y (2000) Scattercast: an architecture for Internet broadcast distribution as an infrastructure service, Ph.D. thesis. University of California, Berkeley
17. Chen H, Guo D, Wu J, Luo X (2006) Theory and network applications of dynamic bloom filters. In: Proc. IEEE INFOCOM, Apr 2006, pp 1–12
18. Choi LK (2006) Overall definition and description of IPTV in the business role model, Tech. report. Telecommunication Standardization Sector Focus Group on IPTV, Jul 2006

19. Cisco Systems, Cisco application-oriented networking. http://www.cisco.com/application/pdf/en/us/guest/products/ps6438/c1650/cdccont_0900aecd802c1f9c.pdf. Accessed Jun 2005
20. Cisco Systems, Cisco application-oriented networking facilitates intelligent radio frequency identification processing at edge. http://www.cisco.com/en/US/products/ps6480/prod_brochure0900aecd8032811f.html. Accessed Jun 2005
21. Cisco Systems, Cisco application-oriented networking streamlines finacial market-data and trade-order latency. http://www.cisco.com/en/US/products/ps6480/prod_brochure0900aecd804b0abe.html. Accessed Jun 2005
22. Clark D, New arch: future generation Internet architecture, Technical Report. http://www.isi.edu/newarch/iDOCS/final.finalreport.pdf. Accessed Jun 2005
23. Coit C, Staniford S, McAlerney J (2001) Towards faster string for intrusion detection or exceeding the speed of snort. In: Proc. DARPA Information Survivability Conference Exposition II, 2001, pp 367–373
24. Commentz-Walter B (1979) A string matching algorithm fast on the average. In: Proc. ICALP, 1979, pp 118–132
25. Dalal YK, Metcalfe RM (1978) Reverse path forwarding of broadcast packets. Commun ACM 21:1040–1048
26. Deering S, Cheriton D (1990) Multicast routing in datagram internetworks and extended lans. ACM Trans Comp Syst 8:85–110
27. Dharmapurikar S, Krishnamurthy P, Sproull TS, Lockwood JW (2004) Deep packet inspection using parallel bloom filters. IEEE Micro 24:52–61
28. dPacket.org. Introduction to deep packet inspection/processing. https://www.dpacket.org/introduction-deep-packet-inspection-processing. Accessed Jan 2008
29. dPacket.org. Digging deeper into deep packet inspection (DPI) https://www.dpacket.org/articles/digging-deeper-deep-packet-inspection-dpi. Accessed Jan 2008
30. Ermolinskiy A, Ratnasamy S, Shenker S (2006) Revisiting IP multicast. In: Proc. ACM SIGCOMM, 2006, pp 15–26
31. Fdida S, Costa L, Duarte O (2006) Incremental service deployment using the hop-by-hop multicast routing protocol. IEEE/ACM Trans Networking 14:543–556
32. Fenner W (1997) Internet group management protocol, version 2. IETF RFC 2236, Nov 1997. http://www.ietf.org/rfc/rfc2236.txt
33. Francis P, Yoid, your own Internet distribution. Yoid project pages. (http://www.icir.org/yoid). Accessed Sept 2000
34. Gribble SD, Lee DC, Polito A, Fox A, Goldberg I, Brewer EA (1998) Experience with top gun wingman: a proxy-based graphical web browser for the 3com palmpilo. In: Proc. Middleware, Sept 1998
35. Guo H, Lo K, Qian Y, Li J (2009) Peer-to-peer live video distribution under heterogeneous bandwidth constraints. IEEE Trans Parallel Distr Syst 20:233–245
36. Helmy A, Thaler D, Deering S, Handley M, Wei L, Estrin D, Farinacci D (1998) Protocol independent multicast-sparse mode (pim-sm): protocol specification. IETF RFC 2362. http://www.ietf.org/rfc/rfc2362.txt
37. Holbrook HW, Cheriton DR (1999) IP multicast channels: express support for single-source multicast applications. In: Proc. ACM SIGCOMM, Aug 1999, pp 65–78
38. IPTV vs. Internet television: Key differences. http://www.masternewmedia.org/2005/06/04/iptv_vs_internet_television_key.htm. Accessed Sept 2005
39. ITU-T Telecommunication Standardization Sector, Quality of experience requirements for IPTV services, Recommendation ITU-T G.1080, Dec 2008
40. Jamin S, Zhang B, Zhang L (2002) Host multicast: a framework for delivering multicast to end users. In: Proc. IEEE INFOCOM, Jun 2002, pp 1366–1375
41. Johnson KL, Kaashoek MF, O'Toole JW Jr, Frans M, Jannotti J, Gifford DK, James K, Overcast: Reliable multicasting with an overlay network. In: Proc. Fourth Symposium on Operating System Design and Implementation (OSDI), pp 197–212

42. Joseph AD, Katz RH, Zhuang SQ, Zhao BY, Kubiatowicz JD (2011) Bayeux: an architecture for scalable and fault-tolerant wide-area data dissemination. In: Proc. ACM NOSSDAV, Apr 2001, pp 11–20
43. Juniper Networks, DSLAM selection for single-edge IPTV networks, white paper. http://www.juniper.net/us/en/local/pdf/whitepapers/2000189-en.pdf. Accessed Mar 2007
44. Juniper Networks, Introduction to IGMP for IPTV networks: understanding IGMP processing in the broadband access networks, white paper. http://s-tools1.juniper.net/solutions/literature/white_papers/200188.pdf. Accessed Jun 2006
45. Kanakia H, Kalmanek CR, Keshav S (1990) Rate-controlled servers for very high-speed networks. In: Proc. IEEE GLOBECOM, Dec 1990
46. Karger D, Kaashoek MF, Stoica I, Morris R, Balakrishnan H (2003) Chord: a scalable peer-to-peer lookup protocol for Internet applications. IEEE/ACM Trans Networking 11:17–23
47. Karp R, Ratnasamy S, Handley M, Shenker S (2001) Application-level multicast using content-addressable networks. In: Proc. 3rd Int. Workshop on Networked Group Communication (NGC'01), 2001, pp 14–29
48. Kermarrec AM, Castro M, Druschel P, Rowstron AIT (2002) Scribe: a large-scale and decentralized application-level multicast infrastructure. IEEE J Select Areas Commun 20:1489–1499
49. Keshav S, Sharma R (2002) Issues and trends in router design. IEEE Commun Mag 36:144–151
50. Knuth D (1997) The art of computer programming: semi-numerical algorithms, 3rd edn. Addison-Wesley, MA, pp 238–275
51. Liao X, Jin H, Liu Y, Ni LM (2007) Scalable live streaming service based on inter-overlay optimization. IEEE Trans Parallel Distr Syst 18:1663–1674
52. Liebeherr J, Beam TK (1999) Hypercast: a protocol for maintaining multicast group members in a logcial hypercube topology. In: Proc. 1st Int. Workshop on Networked Group Communication (NGC'99), 1999, pp 72–89
53. LSI,   http://www.lsi.com/networking_home/networking_products/tarari_content_processors/index.html. Accessed Sept 2010
54. Lyles B, Kassem H, Diot C, Levine B, Balensiefen D (2000) Deployment issues for IP multicast service and architecture. IEEE Network 14:78–88
55. Mccanne S, Amir E, Zhang H (1995) An application level video gateway. In: Proc. ACM Multimedia'95, Nov 1995, pp 255–265
56. McCanne S, Chawathe Y, Fink SA, Brewer EA (1998) A proxy architecture for reliable multicast in heterogeneous environments. In: Proc. ACM Multimedia, 1998, pp 151–159
57. Meeks S, Brooks C, Mazer MS, Miller J (1995) Application-specific proxy servers as http stream transducers. In: Proc. WWW-4, Dec 1995
58. Mitzenmacher M, Byers J, Luby M, Rege A (1998) A digital fountain approach to reliable distribution of bulk data. In: Proc. ACM SIGCOMM, Oct 1998, pp 56–67
59. Moy J (1994) Multicasting extentions to OSPF, IETF RFC 1584. http://www.ietf.org/rfc/rfc1584.txt
60. Nahas M, Liebeherr J, Si W (2002) Application-layer multicasting with Delaunay triangulation overlays. IEEE J Select Areas Commun 20:1472–1488
61. Nicholas J, Adams A, Siadak W (1998) Protocol independent multicast-dense mode (pim-dm): protocol specification (revised), IETF RFC 3973. http://www.ietf.org/rfc/rfc3973.txt
62. Ng TSE, Stoica I, Zhang H (2000) Reunite: a recursive unicast approach to multicast. In: Proc. IEEE INFOCOM, Mar 2000, pp 1644–1653
63. Partridge C, Waitzman D, Deering S (1988) Distance vector multicasting routing protocol. In: IETF RFC 1075. http://www.ietf.org/rfc/rfc1075.txt
64. Pasley J (2005) How BPEL and SOA are changing Web services development. IEEE Internet Comput 9:60–67
65. Perlman R (1999) Simple multicast: a design for simple, low-overhead multicast. Internet Draft. http://tools.ietf.org/html/draft-perlman-simple-multicast-02

66. Ramakrishnan KK, Srivastava D, Cho TW, Rabinovich M, Zhang Y (2009) Enabling content dissemination using efficient and scalable multicast. In: Proc. IEEE INFOCOM, Apr 2009, pp 1980–1988
67. Rexford J, Subramanian L, Agarwal S, Katz RH (2002) Chracterizing the Internet hierarchy from multiple vantage points. In: Proc. IEEE INFOCOM, Mar 2002, pp 618–627
68. Rothenberg CE, Arianfar S, Jokela P, Zahemszky A, Nikander P (2009) LIPSIN: line speed publish/subscribe inter-networking. In: Proc. ACM SIGCOMM, 2009, pp 195–205
69. Shaikh A, Wang J, Yates J, Zhang Y, Mahimkar A, Ge Z, Zhao Q (2009) Towards automated performance diagnosis in a large IPTV network. In: Proc. ACM SIGCOMM, Aug 2009, pp 231–242
70. Sincoskie WD, Wetherall DJ, Tennenhouse DL, Smith JM, Minden GJ (1997) A survey of active network research. IEEE Commun Mag 35:80–86
71. Thaler D, Alaettinoglu C, Estrin D, Kumar S, Radoslavov P, Handley M (2000) The MASC/BGMP architecture for inter-domain multicast routing. IEEE Network 14:10–20
72. Tian X, Cheng Y, Ren K, Liu B (2008) Multicast with an application-oriented networking (AON) approach. In Proc. IEEE ICC
73. Tian X, Cheng Y, Shen S (2012) DAZA: a fast channel zapping scheme for IPTV facilitated by destination-oriented multicast. IEEE Trans Parallel Distr Syst
74. Turner J, Waldvogel M, Varghese G, Plattner B (1997) Scalable high speed IP routing lookups. In: Proc. ACM SIGCOMM, Oct 1997, pp 25–36
75. Verma D, Pendarakis D, Shi S, Waldvogel M (2001) ALMI: an application level multicast infrastructure. In: Proc. 3rd conference on USENIX Symposium on Internet Technologies and Systems (USITS), Mar 2001, pp 49–60
76. VerSteeg B, Begen A, VanCaenegem T, Vax Z (2010) Unicast-based rapid acquisition of multicast RTP sessions. Internet-Draft draft-ietf-avt-rapid-acquisition-for-rtp-17, Nov 2010
77. De Vleeschauwer D, Degrande N, Laevens K, Sharpe R (2008) Increasing the user perceived quality for IPTV services. IEEE Commun Mag 46:94–100
78. Wetherall D, Borriello G, Jain S, Mahajan R, Gribble SD (2002) Scalable self-organizing overlays, Tech. report. University of Washington, Feb 2002. http://citeseerx.ist.psu.edu/viewdoc/summary?doi=10.1.1.89.7315
79. Wu S, Manber U (1994) A fast algorithm for multi-pattern searching, Technical Report TR-94-17. Department of Computer Science, University of Arizona. http://webglimpse.net/pubs/TR94-17.pdf
80. Yang Y, Wang J, Yang M (2008) A Service-Centric Multicast Architecture and Routing Protocol. IEEE Trans Parallel Distr Syst 19:35–51
81. Zhang H, Chu Y, Rao SG (2000) A case for end system multicast. In: Proc. ACM SIGMETRICS, Jun 2000, pp 1–12

Chapter 3
Application-Oriented Multicast

Abstract This chapter presents the basic design of application-oriented multicast (AOM). The service model of AOM is introduced in Sect. 3.1, which improves the scalability of IP multicast. However, several practical design issues have to be considered when realizing the service model over the practical Internet, which are described in Sect. 3.2. The implementation of AOM presented in Sect. 3.3 resorts to the Bloom filter technique to limit the explicit addressing overhead and utilizes the RPF concept to accommodate the longest prefix matching and route aggregation in the Internet, where the first two practical design issues are handled. With the Bloom-filter-based implementation of AOM, receivers send membership updating messages to the data source to construct a reverse SPT to facilitate the data forwarding; moreover, all the multicast membership and addressing information traversing the network is encoded with Bloom filters for low storage and bandwidth overhead. In Sect. 3.4, we describe some complex and detailed situations in the Bloom-filter-based AOM. A theoretical analysis of the Bloom filter performance is given in Sect. 3.5. The last two sections present a comprehensive simulation results and summarize the chapter, respectively.

3.1 AOM Service Model[1]

In the AOM service model, the receiver notifies the source node of the groups it is interested in, with which the source node knows the members for each group it provisions. The multicasting packet carries these members' addresses so that each router can retrieve the addresses and leverage the unicast routing table to compute necessary copies and output interfaces. Correspondingly, the model comprises two components: *membership management* and *forwarding protocol*. We focus on the inter-domain multicast in this thesis, as the scalability is our main concern; the AOM operations within a domain is briefly discussed in Sect. 3.4.4.

---

[1] ©[2008] IEEE. Portions reprinted, with permission, from [15].

X. Tian and Y. Cheng, *Scalable Multicasting over Next-Generation Internet: Design, Analysis and Applications*, DOI 10.1007/978-1-4614-0152-0_3,
© Springer Science+Business Media New York 2013

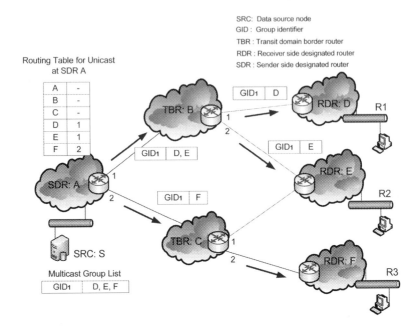

Fig. 3.1 AOM forwarding

3.1.1 Membership Management

For membership management, the border router of a stub autonomous system (AS) domain is selected as a *designated router*. We use RDR and SDR to denote the DR of a receiver-side AS domain and that of a source-side AS domain, respectively. The membership management functionalities of the RDR and SDR are different.

The RDR basically needs to implement the internet group management protocol (IGMP) [12] to discover the active groups, having at least one member host, within its domain. When discovering a new active group, the RDR is triggered to send *membership updating messages* (MUMs) to the SRC in the format as (RDR: GID_1, GID_2, ..., GID_n), where RDR represents a domain prefix and GID represents the group identifier. The MUMs will be delivered along the shortest path between the RDR and the SRC, determined by the unicast routing table.

The SRC aggregates the MUMs it received and maintains a *multicast group list* (MGL). For each group provisioned by the source, the MGL establishes a record in the format as (GID: RDR_1, RDR_2, ..., RDR_n), where each RDR again indicates a domain prefix. When the SRC multicast data over a certain group, it will insert the corresponding MGL into the packet as the destination information in the format of a shim header. The multicast packets are then forwarded to the SDR for inter-domain multicasting.

We use an example as shown in Fig. 3.1 to illustrate the membership management. The active multicast group has a single data source and three active members

R1,R2, and *R3*. The SDR is router *A*, and three RDRs are routers *D*, *E*, and *F*, respectively. Each RDR maintains a GHL, with the GHL record in RDR *F* is illustrated in Fig. 3.1. The group joining messages from the three RDRs will be propagated via unicast to the SDR *A*, where the information is aggregated to a MGL as shown in the figure.

3.1.2 Multicast Forwarding Protocol

The multicast forwarding is facilitated by the AON technique. At each AON router, the normal/AON flag bit and the AON module classifier in the payload will direct the multicast packets to the multicast AON module.

When receiving a multicast packet, each router will extract the MGL record from the packet and compute packet copies and corresponding output interfaces. Specifically, an AOM router performs the following processing. First, check the unicast routing table to determine the output interface for each destination listed in the MGL of the packet and aggregate destinations with the same output interface into a set. Second, replicate the packet for each unique interface found in the first step. Third, update the MGL of each packet copy with the aggregated set yielded in the first step, so that the packet copy for a given interface contains only the destinations that can be reached via this interface. By removing unnecessary destinations from the MGL record, the downstream router will not generate unnecessary packet copies for those destinations that have been delivered over other sibling subtrees. Each router will execute the same operations of aggregation, replication, and MGL record updating, until one multicast packet reaches an RDR.

Consider the example shown in Fig. 3.1; we term the AS domain(s) that connect source and destination stub domains as *transit* domain(s). The IP routing table of the SDR *A* tells that the output interface 1 is on the path to both RDR *D* and *E* so that only one copy is necessary to be forwarded via interface 1. The IP routing table also shows that another copy should be forwarded via interface 2 to reach RDR *F*. When the input multicast packet is replicated and put onto each output interface, the MGL record attached with each copy is updated correspondingly to include only the destination RDRs that can be reached via that interface. For example, the MGL record in the packet delivered over A's interface 1 include only RDRs *D* and *E*. The multicast module at each AON router will execute the same operations of aggregation, replication, and MGL record update, until one multicast packet reaches an RDR.

The service model of AOM multicast mechanism has two important characteristics: First, all the routers involved in the multicast forwarding, other than the DRs, do not need to maintain any states regarding multicasting. The forwarding complexity is totally independent of the number of groups to be supported, resulting in desirable scalability. Second, the membership management component and the multicast forwarding component are completely decoupled. This property enables the flexibility to develop heterogenous application-specific membership management schemes over the same multicast forwarding protocol.

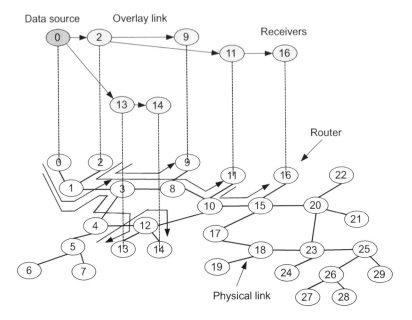

Fig. 3.2 Simulation topology

3.1.3 Service Model Level Evaluation: AOM, IP, and Overlay

In this section, we present some ns-2 simulation results to demonstrate the performance of the proposed AON multicast scheme. For convenience, the illustration figures of the simulations results are generated with Matlab. The network topology for simulation is given in Fig. 3.2, which is similar to that used in [8]. As the essence of the proposed AON multicast is an AON-based forwarding protocol, we here focus on examining the forwarding performance, assuming all memberships already established. Thus, all the nodes in Fig. 3.2 represent routers, and we simulate the multicasting from SDR to RDRs. All the link capacity is set as 1 Mbps, and UDP traffic is used in all the simulations.

We compare the AON multicast with the IP multicast, both in *dense mode* (DM) and *sparse mode* (SM), and the overlay multicast. The IP-DM multicast implements source-based trees with reverse path forwarding (RPF) and pruning, similar in spirit to DVMRP [63]; the IP-SM multicast employs a core-based approach, constructing a tree rooted in a selected *rendezvous point* (RP) [36]. For overlay multicast, two overlay trees are constructed according to a random scheme [6, 7], as shown in Fig. 3.3. For each multicast mechanism, we simulate scenarios with different group sizes (i.e., number of RDRs). With node 0 being the SDR, the group of size 4 includes nodes $\{2, 6, 9, 13\}$. The node sets $\{7, 11, 14, 16\}$, $\{17, 18, 20, 23\}$, $\{19, 21, 24, 25\}$, and $\{22, 27, 28, 29\}$ will be added in turnto form the groups of size 8–20, respectively.

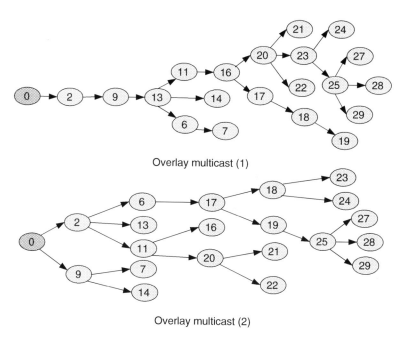

Overlay multicast (1)

Overlay multicast (2)

Fig. 3.3 Two overlay multicast trees

A common measure to demonstrate the multicast bandwidth efficiency is *link cost* [17], which is defined as the total number of physical links that a multicast tree passes for data delivery. For example, the link cost for the overlay multicast tree illustrated in Fig. 3.2 is $2+5+4+5+3+2 = 21$. In the following, we further consider the measures of *bandwidth cost percentage* (BCP), *group receiving rate* (GRR), and *group average delay* (GAD) to examine the multicast performance in depth.

Bandwidth Cost Percentage. The BCP is defined as

$$\text{BCP} = \frac{T}{C \cdot D} \times 100\%, \tag{3.1}$$

where T denotes the total number of bits traversing the physical network, C the total network capacity (i.e., the summation of all link capacities), and D the simulation duration. BCP is a measure of the bandwidth cost of multicast schemes. It is obvious that higher link cost will result in higher BCP.

The BCPs for different multicast schemes are shown in Fig. 3.4. We can observe that, for all kinds of group size, IP-DM multicast and AON multicast have the similar performance, lower than other schemes. It is not surprising that the overlay multicast incurs the highest bandwidth cost due to redundant traffic. Moreover, we can see that the bandwidth cost of the overlay multicast heavily depends on the structure of the

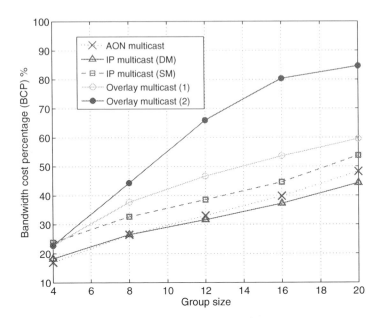

Fig. 3.4 Bandwidth cost percentage

multicast tree. The BCPs of all multicast schemes increase along with the group size, as the multicast will incur more traffic over the network when there are more receivers.

The reason for the bandwidth efficiency of AON multicast is that the AON forwarding protocol in fact implicitly establish a source-based tree from the IP routing table, which is the same as the tree constructed by the RPF used in the IP-DM multicast. Generally, the BCP of AON multicast is slightly larger than that of the IP-DM multicast, due to the bandwidth overhead induced by carrying AON and multicast address information with each packet. Moreover, the bandwidth overhead will bring larger cost when group size becomes larger, as demonstrated in Fig. 3.4. The exceptional case is for group size 4, where AON multicast achieves the lowest BCP. The reason is that the the tree pruning-messages involved in the IP-DM multicast also lead to bandwidth overhead, which exceeds that generated by the AON multicast under small group size.

In the IP-SM multicast scheme, we select node 15 as the RP according to a topology-based RP selection policy [18]; the RP has the smallest average distance (in terms of the number of hops) to other nodes in the network. In IP-SM multicast, data packets are first unicast from the sender to the RP and then multicast over the tree. The traffic between the sender and the RP can be interpreted as bandwidth cost, leading to a higher BCP compared to IP-DM multicast and the AON multicast.

Group Receiving Rate. The GRR is defined as the summation of the data receiving rate at each RDR associated with the group, in terms of bits per second. In this

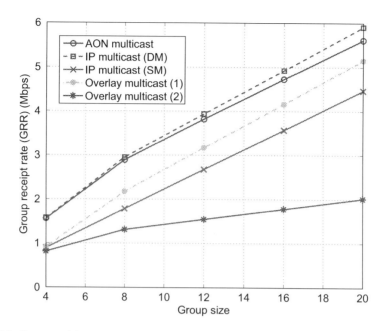

Fig. 3.5 Group receiving rate

experiment, we set capacities of link $(1, 3)$ and $(10, 15)$ to be a quarter of other links, with the objective to investigate the impact of bottleneck links on the performance of different multicast schemes.

Figure 3.5 depicts the "GRR versus group size" curves generated from simulations. In all the scenarios, the source node sends out data at the same rate, so the different GRRs achieved by different multicast schemes are due to the packet loss over the bottleneck links. Since the IP-DM multicast and the AON multicast schemes do not generate any redundant traffic, under these two schemes, the bottleneck congestion and associated packet loss are largely avoided, resulting in good throughput. The GRRs of two overly multicast schemes are much smaller compared to IP-DM and AON schemes, as the redundant traffic incurred by the overlay results in more severe congestions at the two bottleneck links and therefore higher packet loss. One interesting observation is that the IP-SM multicast achieves a GRR even smaller than that of the overlay multicast (1). The reason is that the two bottleneck links are on the path between the source node and RP; therefore, the links seriously impact the throughput of the whole group. It is not difficult to see that such bottleneck impact applying through the RP will take effect independent of the group size; such a fact is indeed demonstrated in Fig. 3.5, where the GRR of the IP-SM multicast is strictly linear with the group size. This kind of perfect linear relationship is not available in all the other cases, where the packet loss is only due to a portion of the tree branches that pass the bottleneck links.

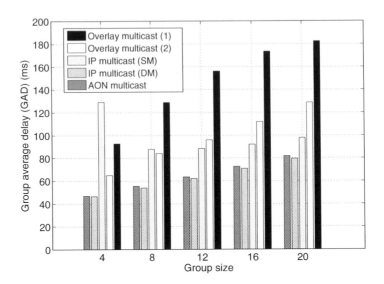

Fig. 3.6   Group average delay

Group Average Delay.   The GAD is defined as

$$\text{GAD} = \frac{\sum_{i=1}^{N} d_i}{N},\qquad(3.2)$$

where d_i represents the delay from the data source to each group member (which is an RDR in our simulations) and N is the group size. The GAD can reflect the timing performance of a multicast scheme, which is important to real-time applications.

We show the results for GAD in Fig. 3.6. The GAD of IP-SM multicast has a different behavior from that of other multicast schemes, which is decreasing versus the group size first and then maintains relatively stable. Such a GAD behavior is determined by the position of RP and the distribution of group members. The first four group members {7, 11, 14, 16} have a relative large distance from the RP node 15, so the GAD is high at first. The later joining group members are closer around the RD, and they flatten out the GAD according to (3.2).

All the other multicast schemes adopt a tree rooted at the source node 0. The GAD grows versus the group size since the later joining nodes are further away from the source node, leading to a larger delay. We can also see from Fig. 3.6 that bandwidth efficient multicast schemes also have better delay performance. The reason is that if a multicast scheme incurs redundant or overhead traffic, the higher traffic load normally will lead to a longer queue at each output interface. Figure 3.6 shows that the bandwidth efficient IP-DM and AON multicast schemes have a significant advantage in the delay performance too. The AON multicast incurs a higher delay over IP-DM. This is because the processing overhead in the AON multicast scheme is longer than that in the IP multicast case; particularly in the current single-group scenario, the multicast routing table has only one entry. We expect that when multiple groups exist, the AON multicast will show advantages over the IP multicast due to its group-independent forwarding design. One of our ongoing work is to examine the performance of the AON multicast scheme over a practical topology supporting a large number of groups.

3.2 Practical Design Issues[2]

Limit the Explicit Addressing Overhead. In the prototype AOM service model, the forwarding complexity is totally independent of the number of groups to be supported, resulting in desirable scalability. Nevertheless, considerable bandwidth overhead could be incurred when there are a large number of receivers (RDRs) for each group: The MGL in the packet becomes impractically long, and the number of receivers that can be supported is constrained by the packet header size.

Accommodate Longest Prefix Matching and Route Aggregation. A possible solution of limiting the explicit addressing overhead is to encode the MGL into a Bloom filter [9] which allows the information to be compressed. However, the Bloom-filter-based design needs to support the features of Internet. Normally, Internet routers apply the longest prefix matching and route aggregation schemes to control the size of the unicast routing table, thus the same destination network may be represented with different network prefixes in different routers. Since the Bloom filter only supports exact query, it is possible that the destination RDR prefixes encoded in the Bloom filter cannot match any forwarding entry stored in an SDR/TBR. Instead of direct utilizing the unicast routing table, there is a need to establish the forwarding states that can recognize the Bloom-filter-formatted MGL along the data delivery path.

Work with the Asymmetric Inter-Domain Routing. Most of the multicast protocols in practice [1, 10] establish the forwarding states when the joining request is delivered from the receiver to the source node (or RP) and then forward the data packets along the path that is reverse to the joining path. The purpose is actually using the multicast forwarding states to label a source-based reverse SPT; however, constructing the reverse SPT requires the symmetric routing environment: The path from the source to a receiver follows the same path used to go from the receiver to the source. Unfortunately, the inter-domain routing is usually asymmetric for the administrative reasons [10]. When designing the AOM, we also have to consider the effect of asymmetric routing on the protocol so that the proposed protocol can be applied in the practical Internet.

Eliminate Loops Caused by False Positive. The Bloom filter incurs false positive, which means that an element not encoded in the Bloom filter can be falsely detected. In some subtle cases, the false positive can result in forwarding loops, which

---

could cause the partial breakdown in the network. AOM should have the ability to eliminate the loops caused by the Bloom filter false positive, while the cost of the ability should be constrained.

Support Incremental Deployability. The evolution to next-generation Internet needs continuing efforts; thus, it is impractical to upgrade all routers to be aware of AOM simultaneously. AOM needs to be incrementally deployable: it should be able to work in the network, in which only a small fraction of routers are aware of AOM protocol while others are legacy routers. The correctness of AOM should not be affected but some efficiency may be lost.

3.3   Bloom-Filter Based Implementation of AOM[3]

This section presents a streamlined Bloom-filter-based design to achieve the AOM with reasonable cost. We take the assumption of symmetric unicast routing for the convenience of demonstration and discuss later how to extend AOM for operating in asymmetric routing scenarios.

3.3.1   Bloom Filter Data Structure

We are to describe the Bloom-filter-based design of AOM according to the upstream procedure (i.e., states establishment) and downstream procedure (i.e., data forwarding), as illustrated in Fig. 3.7, where Bloom filters are illustrated as shadowed areas.

The left side of Fig. 3.7 shows how forwarding states are established by joining MUMs. To reduce the bandwidth overhead for membership updating, the list of active groups in the MUM is encoded with a *group Bloom filter* (GRP_BF). When an MUM reaches an upstream TBR/SDR router, the router will retrieve the RDR prefix and store it as a local forwarding state at the output interface corresponding to the MUM incoming interface; the local status will later be used for RPF. By continuously observing the MUMs, each related interface of the TBR/SDR will memorize all the destination domains that can be reached through it, and the reverse SPT from the SRC to subscribing RDRs is constructed. At an output interface, each RDR is stored as a separate Bloom filter, termed as *interface RDR Bloom filter* (IRDR_BF), which will be used to facilitate multicast forwarding.

The upstream MUMs will finally reach the SRC node, and each message will be stored as a record of the MUM table. The SRC node should have a *local channel list* indicating the multicast groups it provisions. By checking each GID

---

[3] ©[2009] IEEE. Portions reprinted, with permission from [16].

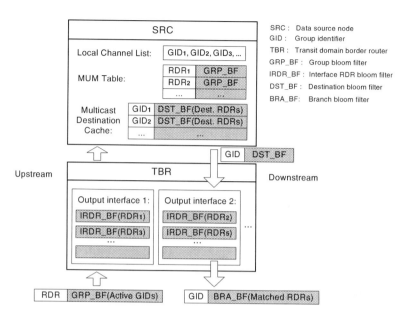

Fig. 3.7 Bloom filter data structure for AOM

against the MUM table and identifying the matched GRP_BF, the SRC can detect the destination prefixes for a given group. The destinations information under the group ID will be encoded into a *destination Bloom filter* (DST_BF) and stored into the multicast destination cache. Note that the DST_BF in fact encodes the MGL according to the AOM service model.

The right side of Fig. 3.7 illustrates how multicast packets are forwarded. At the SRC node, the DST_BF for a group will be inserted as the destination information into each multicast packet. In the downstream data forwarding process, each router generally executes the same operations of aggregation, replication, and MGL record updating as introduced in the AOM service model. The only difference is that these operations are conducted with Bloom filters in both the packet and the router. Specifically, each TBR/SDR compares the packet's DST_BF with IRDR_BFs at each interface. A packet replica is generated and dispatched along the interface if the DST_BF and the IRDR_BFs installed at the interface have any element matched. The subset of matched prefixes associated with each output interface is then reencoded into the *branch Bloom filter* (BRA_BF). The BRA_BF will be inserted into the packet replica delivered through that interface, serving as the destination information DST_BF for further downstream forwarding.

The forwarding states stored at the router are destination specific and totally independent of the number of groups passing through the router. Compared with IP multicast, the number of required forwarding states could be significantly reduced since AOM stores only one state on each related intermediate router for each subscriber domain, but each subscriber domain may join in tens of

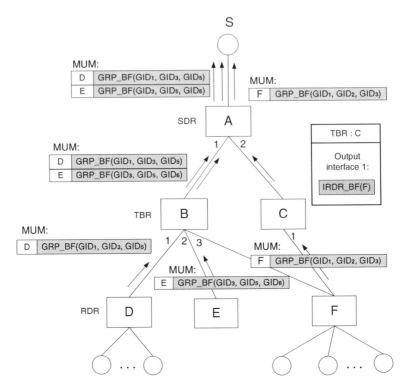

Fig. 3.8 Membership updating process of AOM

thousands of groups, and each group needs a state on the related router under IP multicast. The downstream procedure actually follows the *RPF concept: the data packet is forwarded along the path that is reverse to the MUM joining path.* The states installed by MUMs explicitly construct an SRC-based reverse SPT, and the forwarding procedure does not need to utilize the unicast routing information to decide output interfaces; thus, the longest prefix matching and route aggregation schemes can be accommodated in AOM.

In the following, we are to present the implementation details of AOM, including group joining process, MUM processing at the SRC node, downstream data forwarding, and group leaving process, by studying an example topology as shown in Fig. 3.8.

3.3.2   Group Joining Process

The membership updating procedure is illustrated in Fig. 3.8. The MUMs are sent from RDRs to the source by leveraging the unicast mechanism. The updating is normally used to register the new active groups and unregister the inactive groups. We will discuss the group leaving process in a separate subsection.

In Fig. 3.8, the path *F-C-A* shows the joining process of RDR *F*. When *F* detects active groups 1, 2, and 3, it generates the MUM in the Bloom filter format as [F: GRP_BF(GID$_1$, GID$_2$, GID$_3$)] and then sends the MUM toward the source node *S*. When the MUM reaches the upstream node *C* through interface 1, node *C* will retrieve the prefix associated with RDR *F* and put *F* into a separate Bloom filter IRDR_BF at interface 1 as the multicast forwarding state for reverse path routing. Node *C* continues forwarding the MUM to *A*, and *A* implements local status updating in a similar way. When the MUM reaches the SRC node *S*, it will be stored as a record into the MUM table.

Reducing the Bandwidth Overhead. The MUM updating may result in considerable bandwidth overhead due to two reasons: (1) The number of active groups in a domain may be large, which will require a big size GRP_BF to maintain the false-positive performance, and (2) groups may turn on/off frequently, leading to frequent MUMs. A bandwidth-efficient MUM updating approach is suggested in [9]. An RDR sends an MUM to the SRC each time when a group appears or departs, but the updates are conveyed as the set of GRP_BF bit positions to be set/reset. The source node will use the position information to update the MUM record for the corresponding RDR. For example, if the GRP_BF uses 5 hash functions and bit positions are represented as 24-bits values, each position update MUM only consumes 15 bytes [9].

GRP_BF Design. When a source node searches the destination domains for a multicast group by checking the MUM table, a certain RDR may be mistaken as the destination due to the false-positive property of the Bloom filter. Since a false-positive destination will result in redundant traffic, the Bloom filter GRP_BF should be properly designed. Since the false positives are due to the nature of Bloom filter and cannot be totally avoided, each false positive usually results in a redundant traffic filter being installed at the upstream domain. Let *A* denote the number of multicast groups provisioned by a given source node and *G* denote the number of active groups associated with that source node subscribed by an RDR domain. If an RDR destination domain is allowed *f* redundant traffic filters, the target false-positive ratio for GRP_BF can be set as $\min(1, f/(A-G))$ [9]. Such a choice follows the observation that a false positive can only be triggered by one of the $A-G$ groups. The target false-positive ratio is used to guide the design of the GRP_BF filter size.

3.3.3 MUM Processing at the SRC

The MUMs initiated from RDRs will be processed at the SRC. The SRC node has the list of channels it is offering. By checking all the GIDs in the list against the MUM table, the SRC can detect which groups have been requested and the

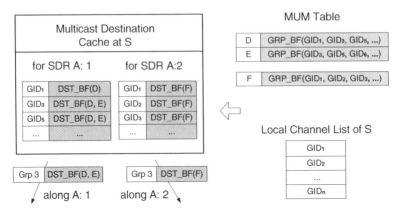

Fig. 3.9 Message processing at data source

associated destination domains. For a certain GID, all the RDRs associated with the GRP_BFs confirming the checking will be encoded into a DST_BF and then inserted into the multicast destination cache, indexed by the GID. In principle, the operations of destination search only execute for the first data packet of a group, and the catched DST_BF will be applied to later packets. In practice, the SRC needs to periodically check the MUM table and refresh the destination cache so that the destination updating information carried by MUMs can be incorporated timely. Another implementation detail is that the SRC and RDRs should agree on the set of hash functions used by the GRP_BF Bloom filters, which are generated by the RDRs but used for information retrieving by the SRC.

Figure 3.9 illustrates the operation at S of Fig. 3.8. Take group 3 as an example. Since GID_3 is contained in the three GRP_BFs indexed by D, E, and F, the operation checking the MUM table will generate a record for the multicast destination cache as [GID_3: DST_BF(D, E. F)]. Figure 3.9 shows the details that the SRC in fact generates two separate DST_BFs based on the MUM incoming interfaces at the SDR; the interface information can be attached to the MUM by the SDR. It can be seen that the two DST_BFs contain the destination information along the two subtrees rooted at the SRC. For each multicast data packet, the SRC will generate two copies with the two DST_BFs attached correspondingly. Such a splitting technique can efficiently reduce the number of items put into the Bloom filter for better false-positive performance.

It is noteworthy that, in our scheme, the multicast packet preparation is taken care of by the source node rather than the SDR. Such an implementation will considerably improve the scalability when a source domain contains a large number of multicast source nodes, compared with the FRM approach [9] which implements all the group-specific processing at the SDR node.

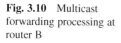

Fig. 3.10 Multicast forwarding processing at router B

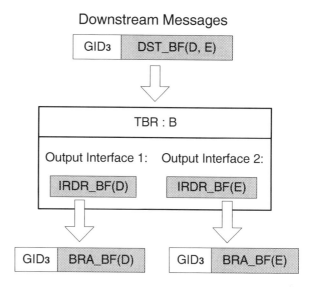

3.3.4 *Downstream Data Forwarding*

The downstream message processing is illustrated in Fig. 3.10, which captures the scenario that a multicast packet for group 3 is at router *B*. Note that although the DST_BF carries the destination information, we cannot just directly check the unicast forwarding entry against DST_BF to determine the output interfaces. The reason is that the unicast routing table normally applies route aggregation and longest prefix match, particularly in an inter-domain context, where the accurate RDR address information is not available to query the DST_BF. The local IRDR_BFs stored at each interface are important components to enable the AOM.

The AOM routing/forwarding is implemented through comparing the DST_BF against the local IRDR_BFs, which requires that the SRC node and the network agree on the size of the Bloom filter size and the set of hash functions involved. For example, suppose 5 hash functions are used; if the comparison between the DST_BF and a IRDR_BF confirms 5 matched "1" bits, we define that a matched IRDR_BF is identified. Each interface containing matched IRDR_BF(s) represents a tree brand according to the RPF concept, over which a separate copy of the data packet will be delivered. Moreover, a BRA_BF will be generated by aggregating all the matched IRDR_BFs (through an "OR" operation) over that interface. The BRA_BF will be used as DST_BF for further downstream forwarding.

We would like to emphasize that replacing the incoming DST_BF with the corresponding BRA_BF over each output interface achieves the aggregation, replication, and MGL record-updating operations defined in the AOM service model, which will successfully avoid redundant traffic according to Sect. 3.1.1. In Fig. 3.8, the incoming DST_BF contains both *D* and *E* as destinations. With interfaces 1 and 2 are determined as forwarding interfaces, the BRA_BF over each interface removes the destination processed by the other branch. It would be remiss not to mention that matching the DST_BF against the IRDR_BFs may generate false positives, which will be analyzed in detail in Sect. 3.5.2.

3.3.5 Group Leaving Process

When the number of users for a certain group has fell to zero, the RDR will trigger an MUM to notify the SRC of the change, similar to the MUM updating in the joining process. The difference is that the bit positions indicated by the MUM will be reset in the corresponding GRP_BF record in the SRC.

Consider the scenario in Fig. 3.8. If the group 1 in F's domain has no active registered users, RDR F will not expect any traffic for group 1. An updating MUM will reach the SRC and change the related GRP_BF record to [F: GRP_BF(GID$_2$, GID$_3$)]. In next operation of checking the MUM table, the new destination information will be passed to the multicast destination cache. Specifically, F will be removed from the destination cache record for group 1 and therefore stops forwarding group 1 packets to F.

When there is no active group in a receiver domain, the domain's RDR may leave the multicast tree. In the aforementioned example, supposing that F is leaving, it sends the MUM (F: NULL) to S. Along the upstream path of the MUM, related routers will delete the local IRDR_BF storing F. After S gets this "NULL" MUM, it will delete the MUM record indexed by F and update the destination cache through the coming MUM table checking. We can see that a router with IRDR_BF(F) removed from the local states will immediately stop forwarding traffic to F.

3.4 Discussions on Implementation Details

3.4.1 Services Decoupled from Routing

The architecture design of AOM described in Sect. 3.1 completely decouples the membership management component and the multicast forwarding component. This property also holds in the Bloom-filter-based design. Consider the scenario in Fig. 3.8. Forwarding component uses RDR and IRDR_BF information in both MUMs and local routers to find the reverse shortest paths from SDR to RDRs. Moreover, group IDs are only used to label groups at SRC and RDR; it is not according to group IDs that routers construct delivery tree or make forwarding decisions. How to identify a group can be totally decided by the SRC. Therefore, it is not necessary to use Class D IP address to distinguish one group from another, as long as the group ID is unique at the SRC.

A group can be identified by a tuple of the SRC IP address and a locally unique channel number, represented as $< SRC\_IP, Channel\_Num >$. Comparing with existing multicast address allocation mechanisms, this localized address allocation method avoids the collision happens in the random allocation scheme. It does not only averts the scalability problem as in static allocation proposal but also save the complex configurations as in MASC. The performance evaluation in the next section will reveal that the forwarding complexity is totally independent from the number of active groups in the system. The proposed address allocation scheme can well suit for applications such as IPTV or videoconferencing which require active dynamic membership switch or supporting large numbers of small groups.

Channel discovery can be implemented in a number of manners, including via web pages or session description broadcast programs, etc. This is similar to what have been used for traditional multicast applications.

3.4.2 Asymmetric Routing Scenario

The AOM implementations described so far are under the assumption of symmetric routing, in fact, it can be readily extended to asymmetric routing scenarios. As the inter-domain routing is normally policy based [10], a border router will be aware of the asymmetric routing policy applied to a given destination domain. In case an SDR receives an MUM in the asymmetric case, in addition to forwarding the MUM to the source node, the SDR will further pass the MUM along the downstream path to the destination RDR. The local IRDR_BFs will be installed at the corresponding output interfaces along the path from SDR to the RDR. The MUM returned back also behaves as a notification signal to the destination RDR that asymmetric routing is the policy and the multicast forwarding states has been established successfully. As an extra step, the RDR then send a new type MUM (F: asymmetric) to the SRC to remove the local IRDR_BFs that had been established under the symmetric assumption. As long as the local IRDR_BF states are correctly established, other AOM implementation details apply to both the symmetric and asymmetric cases.

As shown in Fig. 3.8, a possible asymmetric routing scenario is that the unicast path from F to S, F-C-A-S, is asymmetric to the path from S to F, S-A-B-F. In such a case, the MUM sent by F continues traveling from S to F after it reaches S so that IRDR_BF(F) can be installed at the interface 1 of A and the interface 3 of B.

3.4.3 Supporting the Multisource

AOM can support the multisource scenario. To this end, a counter is installed to monitor the IRDR_BF for the number of joining messages passed through the TBR. Figure 3.11a, b shows how the joining mechanism works in the existence of double

Fig. 3.11 Supporting the multisource scenario

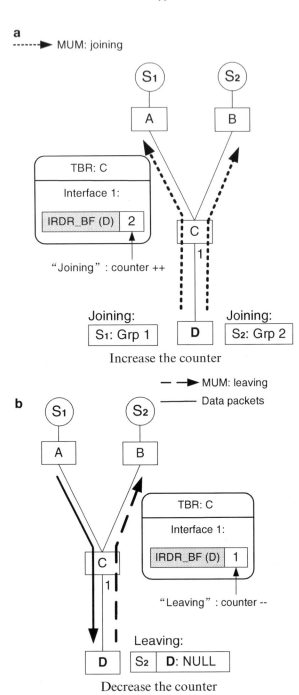

sources. Suppose D wants to join group 1 and group 2 at S_1 and S_2, respectively. It just sends two regular MUMs toward the SRC as shown in Fig. 3.11a. When the MUMs are received by C, it can tell that D has sent out two joining messages. C then increases the counter associated with IRDR_BF(D). If D wants to leave the multicast tree rooted at S_2, it sends the MUM (D: NULL) toward S_2. When the message passes through C, the TBR will decrease the counter associated with IRDR_BF(D) as shown in Fig. 3.11b. In this way, the forwarding state IRDR_BF(D) can still be used for the packets sent from S_1. The IRDR_BF will be deleted when the counter equals zero.

We could use two bits in the MUM to distinguish the message's different purposes. Each time an RDR joins to a new SRC, the TBRs along the joining path will increase the counters associated with the corresponding IRDR_BFs; each time an RDR leaves a current SRC, the TBRs along the leaving path will decrease the counters associated with the corresponding IRDR_BFs. The regular membership updating operations (e.g., D may want to join in other groups at S_1 after subscribing to the S_1 with group 1) will not affect the counters in TBRs.

3.4.4 Uniformed Intra-/Inter-domain Solution

The AOM protocol can be applied in a uniformed manner for both intra-domain and inter-domain multicast. For a transit or stub domain, no matter what intra-domain routing protocols are running, the MUMs entering the domain may label the IRDR_BFs not only at the border routers but also at the corresponding core routers within the domain. According to our forwarding design presented above, it can be seen that the downstream packets will find the end-to-end path according to the reverse-path-forwarding operation. As a comparison, the FRM [9] only code the inter-domain multicast tree in the packet header, so it requires additional intra-domain multicast scheme to achieve a complete multicast solution. If a transit domain is not equipped with an intra-domain multicast protocol, then N-unicast or broadcast has to be used to handle the transit-domain FRM traffic, which will lead to significant amount of redundant traffic.

We simplify and represent the AS domain as a border router for demonstration convenience in the previous description. The enlarged view inside a transit domain is illustrated in Fig. 3.12, which helps to understand how AOM protocol can operate within a domain. For domains where the unicast routing is symmetric or the link-state protocols (which enables each node to know the entire domain topology) are running, the AOM protocol can also be applied. In this case, the MUMs entering the domain can somehow correctly install forwarding states based on the RPF concept in such domains, and the downstream packets will find the end-to-end path to cross the domain following our forwarding protocol. In domains dominated by distance-vector-based protocols, each node cannot know the entire topology; thus, installing

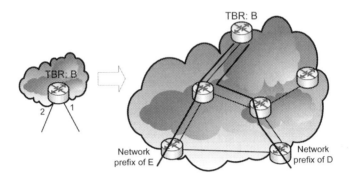

Fig. 3.12 AOM operations within a domain

forwarding states according to the RPF concept could be difficult. Multiple unicast can be used to deliver the inbound AOM traffic to the outbound border routers based on the neighboring domain network prefixes.

3.5 Theoretical Analysis on AOM Forwarding[4]

3.5.1 Loop-Free Forwarding Without Redundant Traffic

The AOM protocol based on the service model presented in Sect. 3.1 ensures loop-free forwarding and incurs no redundant traffic. We have the following theorems.

Theorem 3.1. *The AOM downstream forwarding is free from directed cycles if the following conditions hold: (i) The domains associated with the SDR and RDRs are stub domains of the multicast group under consideration, (ii) the unicast routing in the network is stable, and (iii) the Bloom filter implementation incurs negligible false positive (the false-positive performance is to be analyzed separately).*

Proof. Consider a network supporting multiple multicast groups. We assume that the AOM may lead to directed cycles and then derive the contradictions. Due to the uniformed solution of AOM for both intra- and inter-domain cases, we here give the proof regarding the border routers for convenience. We can see that for a given multicast group, a directed cycle or a forwarding loop can only appear as two cases.

Case 1: The directed cycle takes the form $(TBR_i \rightarrow RDR_j \rightarrow \cdots TBR_k \cdots \rightarrow RDR_l \rightarrow TBR_i)$. In this case, after a packet reaches the destination RDR, it will be further forwarded to other RDRs or TBRs and result in the loop, which obviously conflicts with the condition (i) that the RDRs are stub domains of the multicast group under consideration where further packet delivery for this group should stop.

---

[4]©[2009] IEEE. Portions reprinted, with permission, from [16].

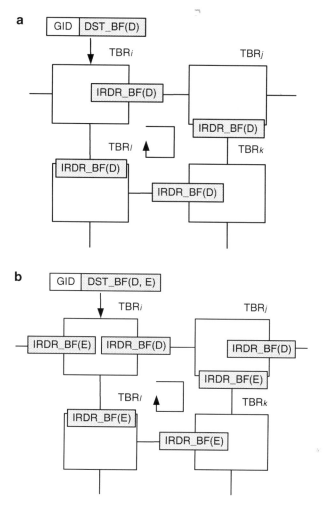

Fig. 3.13 Illustration of directed cycles

Case 2: The directed cycle takes the form (TBR$_i$ → TBR$_j$ → ··· TBR$_l$ → TBR$_i$). In this case, the directed cycle forms among TBRs within the network before reaching the destination domain. This case further includes two subcases.

For the first subcase, as illustrated in Fig. 3.13a, all the interfaces consisting of the forwarding loop are labeled with the same IRDR_BF, which are supposed to be along the path to the destination. According to our reverse-path-forwarding design, the directed cycle in the downstream path implies that the upstream path taken by the MUMs also contains directed cycle. However, in the AOM design, the upstream MUMs exploit the existing unicast path, which is loop-free in stable state, and we get a contradiction.

For the second subcase, as illustrated in Fig. 3.13b, the interfaces consisting of the forwarding loop are labeled with different IRDR_BFs. The IRDR_BF configuration in Fig. 3.13b is possible if the destination domains D and E submit to different groups through different path. According to the AOM design, if two neighboring links (termed as one upstream and one downstream) along a path are labeled with different IRDR_BFs, it means that the node containing the downstream link is a branching node that intersects different paths to different destinations. Note that the AOM protocol incorporates the MGL record-updating operation, that is, after the branching node processing, the destinations associated with other branches will be removed from the DST_BF for those downstream packets. The MGL updating operation ensures that the forwarding loop as shown in Fig. 3.13b is impossible, if condition (iii) holds.

Summarizing case 1 and case 2, the theorem is proved.   □

Theorem 3.2. *The AOM downstream forwarding generates no redundant traffic under the same conditions as applied in Theorem 3.1.*

Proof. Theorem 3.1 has proved that the directed cycle does not exist, so there is no redundant traffic caused by forwarding loops. Consider two additional cases which may potentially generate redundant traffic. Case 1 is that multiple copies of the same packet reach the same RDR from different paths. Case 2 is that some packets reach a TBR but will be dropped due to nonexistence of matching output interfaces at that TBR. According to the reverse-path-forwarding principle of AOM, case 1 implies that some MUMs have labeled more than one path, which is impossible under condition (ii). Moreover, based on the MGL record-updating operation and the condition (iii), we can see that the redundant traffic indicated in case 2 is impossible either.   □

3.5.2 False Positives in Forwarding

False Positive on an Interface. In the AOM forwarding process, bit matching between the in-packet DST_BF and the local IRDR_BF may incur false positives. We can analyze the bit matching in a more general context. Assume two Bloom filters, BF_1 and BF_2 are represented as m-bit arrays and generated by the same k hash functions. BF_1 and BF_2 contain n_1 elements and n_2 elements, respectively.

The bit-matching false positive may happen in three cases: (1) an element in BF_1 but not in BF_2 is positively detected in BF_2; (2) an element in BF_2 but not in BF_1 is positively detected in BF_1; (3) an element neither in BF_1 nor in BF_2 is positively detected in both of them.

The Bloom filter theory [4] tells that the false-positive probability associated with BF_1 and BF_2 are $f_{n_1} = \left(1 - (1 - \frac{1}{m})^{n_1 k}\right)^k$ and $f_{n_2} = \left(1 - (1 - \frac{1}{m})^{n_2 k}\right)^k$, respectively. It is not difficult to see that the total probability for bit-matching false positive due to either case 1 or case 2 can be expressed as

$$f_1 = 1 - (1 - f_{n_2})^{n_1} \cdot (1 - f_{n_1})^{n_2}. \qquad (3.3)$$

The bit-matching false positive due to case 3 is

$$f_2 = f_{n_1} \cdot f_{n_2}. \qquad (3.4)$$

Thus, the total bit-matching false-positive rate is

$$f(n_1, n_2) = f_1 + f_2. \qquad (3.5)$$

In AOM forwarding, the false-positive rate can be computed with n_1 equal to the number of elements in the DST_BF and $n_2 = 1$.

False Positive Along a Path. In AOM, if a false positive, say for destination RDR$_i$, happens at a certain output interface along a path, it will persist until it reaches the destination. The reason is that the MGL updating operation will not remove the RDR$_i$ from the DST_BF, although it is confirmed by false positive, and thus the packet will finish the path labeled by the IRDR_BF(RDR$_i$). Moreover, new false positives may happen in a downstream node due to other RDRs. It is noteworthy that the MGL updating operation has a side benefit to reduce the false positive along a path because the items contained in the DST_BF or BRA_BF continuously become less. We consider that the inter-domain multicast tree can be modeled as a binary tree with height H [5]. For a given destination RDR, an upper bound of the probability that the RDR receives traffic by false positive, denoted as F_p, can be expressed as

$$F_p(H) = \sum_{h=1}^{H} f\left(\frac{n_1}{2^{h-1}}, 1\right), \qquad (3.6)$$

where n_1 is the number of elements contained in the DST_BF generated by the SDR and $\frac{n_1}{2^{h-1}}$ is the number of RDRs in the updated BRA_BFs at the layer-i TBR. We present the specific numerical analysis in Sect. 3.6.3 to show the efficiency of AOM in terms of false positive.

3.6 Performance Evaluation[5]

In this section, we present some ns-2 [14] simulation results and numerical analysis to demonstrate the efficiency of AOM on bandwidth utilization and small false positives.

The network topology for simulation is given in Fig. 3.14, which is widely used in the literature to approximate the US backbone network [13]. In our model, each

---

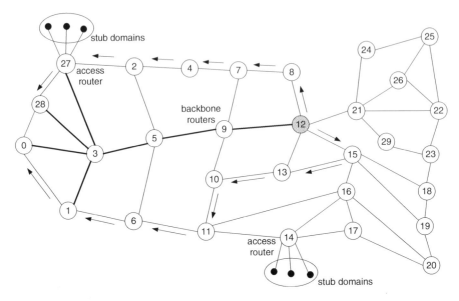

Fig. 3.14 Simulation topology

transit domain is represented as a backbone router or backbone node, and the backbone routers where stub domains are connected are termed as *access routers* or *access nodes*. The multicast source is located at node 12. We simulate multiple multicasting scenarios, where the multicast group involves different numbers of access routers, and evaluate the bandwidth consumption across the backbone network. The scenario involving four access routers includes nodes $\{2,4,6,8\}$. The node sets $\{11,14,17,18\}$, $\{1,19,23,27\}$, $\{0,10,24,25\}$, and $\{13,20,28,29\}$ will be added in turn to form the scenarios involving 8, 12, 16, and 20 access routers, respectively.

3.6.1 Memory Overhead

With AOM, the memory overhead in the source node is mainly on maintaining an MUM table to store the GRP_BF. The GRP_BF size associated with a given receiver domain is designed to maintain a target false-positive rate of $\min(1, f/(A-G))$. We use real BGP table of AS6447 [2] for the numerical analysis. There are 276,058 stub domains, and the percentages for domains with different prefix length p are given. Given the number of users $U(p)$ in a p-prefix domain, if each user joins 100 groups, the number of active groups G in a domain can be determined from the zipfan distribution. Given $f = 10$ and $A = 2^{20}$, the size of GRP_BF associated with a receiver stub domain can then be determined to meet the target false-positive

Table 3.1 Memory overhead comparison

States	Location	When used	Scaling
AOM			
GRP_BFs	SRC	Per group	$O(n_R \cdot g_R)$
DST_BFs	SRC	Per packet	$O(g_A \cdot C(g_A))$
RDR_BFs	SDR,TBR	Per packet	$O(n_R)$
FRM			
GRP_BFs	SDR_s	Per group	$O(n_R^s \cdot g_R^s)$
Cached GRP_BFs	SDR_s	Per packet	$O(g_A^s \cdot T(g_A^s))$
Encoded links	TBR	Per packet	$O(d)$
Grönvall			
EDGE_BFs	TBR	Per packet	$O(d \cdot g_t)$

rate. In the worst case, if there are MUMs from all the 276,058 stub domains to the source node, the total memory for the MUM table is around 2 Gbytes, which is quite affordable under the current memory technology.

The memory overhead at the SDR/TBR is for storing IRDR_BFs at each interface as forwarding states. This part of memory cost is in the order of the number of RDRs passing MUMs through the SDR/TBR. It is suggested to use a 100-bytes shim header within each packet to carry the Bloom filter for multicasting [9]. As AOM implementation incurs Bloom filter match operation, so the local IRDR_BF should also be 100 bytes. The real BGP table [2] suggests that the number of RDRs can be in the order of $O(10^5)$. Then, the IRDR_BF memory overhead at each interface will be in the order of $O(100 \times 10^5) = O(10M)$ bytes, which can easily be accommodated by a line card [9]. It is noteworthy that the local memory overhead at the SDR/TBR is independent of the number of groups and the number of end users being supported.

Table 3.1 summarizes the scaling property of the memory overhead for three Bloom-filter-based multicast schemes. The notations involved are as follows:

- n_R is the number of receiver domains associated with a source node, and g_R is the average number of distinct groups each domain requests.
- g_A is the number of active groups provisioned by the source node, and $C(g_A)$ is the average number of receiver domains for groups offered by the source node.
- n_R^s is the number of receiver domains associated with all the source nodes in the source domain s, and g_R^s is the average number of groups each receiver domain requests from domain s.
- g_A^s is the number of groups with active sources in the domain s, and $T(g_A^s)$ is the average size of the dissemination trees for groups with source in s.
- d is the AS degree.
- g_t is the average number of active groups supported by an interface at a TBR.

We note that the source-routing-based FRM memorizes no multicast routing information at TBRs, while the Grönvall's method maintains the entire routing information at TBRs [4], where a Bloom filter is associated with each interface of the router encoding all the group IDs that should be forwarded along that interface. AOM makes a trade-off between the two approaches. In AOM, the multicast routing/forwarding information is distributed to both packet headers and local states in the router. Compared with FRM, AOM distributes the storage and computational burden at the SDR node to the multiple SRCs within the domain, which eliminates the potential performance bottleneck at the SDR. Moreover, the local forwarding states can alleviate the routing information carried by the packet and thus lead to higher bandwidth efficiency and smaller false-positive rate, which are to be discussed in the following subsections.

3.6.2 Bandwidth Overhead

The AOM bandwidth overhead is incurred by both membership updating and data forwarding. Since we adopt the method suggested by FRM [9] to update the group membership, we here mainly focus on the bandwidth overhead in data forwarding.

The AOM forwarding incurs bandwidth overhead due to two reasons: (1) Each packet needs to carry the destination information, that is, the DST_BF, using a shim header. (2) When the destination entities are too many to be encoded in a single shim header, redundant packet copies have to be incurred, each of which carries a subset of destinations in its own shim header. The similar approach has been adopted by FRM to use multiple packets to carry a big source-routing tree. We fix the size of the shim header ($<10\%$ of the packet size) for AOM and FRM and measure the overhead in terms of the number of packets transmitted.

Total Packet Transmission. The total number of packet transmissions W_x is defined as

$$W_x = T/C, \tag{3.7}$$

where T denotes the total number of transmissions over the whole backbone network within the simulation duration and C the total number of packets generated by the source. W_x represents the average number of transmissions over the whole network to multicast a single packet from the source to all the access routers.

Figure 3.15 shows the values of W_x versus N, the number of access routers involved in multicasting, with different multicast schemes compared. The case $N = 2$ (node 4 and 8 involved) represents a multicast scenario with very sparse node distribution. We observe that, among all kinds of multicast schemes considered, AOM achieves the best performance when the number of access routers N is small or moderate. When N grows large, AOM still performs close to IP multicast. Except for the case of $N = 2$, AOM always outperforms FRM, especially when N is large and the multicast tree has many branches. Moreover, W_x increases along with the number of access routers because each multicast packet has to travel over more links to reach more destinations, which incurs more transmissions across the network.

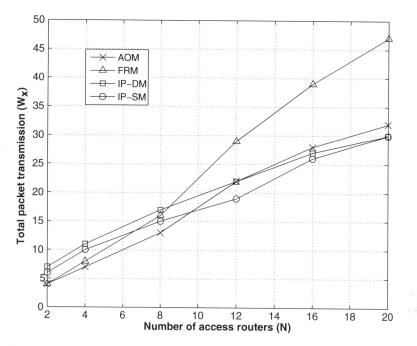

Fig. 3.15 The total number of packet transmissions, W_x

Compared with IP multicast dense mode (IP-DM), AOM in fact implicitly establishes a source-based tree from the IP routing table (referring to Sect. 3.3.3), which is the same as the tree constructed in IP-DM. However, AOM does not use the "broadcast-and-prune" [1] method to construct the multicast tree as used by IP-DM. In the sparse scenarios with small N, the larger values of W_x are due to unnecessary transmissions during the periodical "broadcast-and-prune" operation. When N becomes large, IP-DM becomes favorable since multicast in a densely populated network approaches broadcast and few number of broadcast packets are wasted.

The reason for better performance of AOM over FRM is that AOM needs to encode less elements than FRM does [9]. For example, consider the case that node 12 multicasts data to access routers $\{0, 1, 27, 28\}$; the multicast tree is highlighted in Fig. 3.14 with thick lines. For illustration purpose, consider that the shim header can only encode four elements. In such a scenario, one shim header can encode all the four destinations under AOM. Under FRM, the 7-branch tree needs to be encoded, which exceeds the capacity of one shim header. Therefore, four shim headers over four packets have to be used, each containing the tree branches to one of the destinations. Three of such four packets are counted as redundant traffic compared to AOM. For the case of $N = 2$, the shim header is capable of containing the entire forwarding information for both AOM and FRM, so they have the same W_x.

Table 3.2 Link packet transmission distribution

	AOM 29 links		FRM 29 links		IP-DM 50 links		IP-SM 29 links	
	%	L_x	%	L_x	%	L_x	%	L_x
			5.0	4				
			10.0	3				
$N=12$	10.0	2	10.0	2	36.0	1	11.1	2
	90.0	1	75.0	1	64.0	<1	88.9	1
			3.4	7				
			3.4	5				
$N=20$			3.4	4				
			6.9	3				
	10.3	2	3.4	2	54.0	1	7.1	2
	89.7	1	79.5	1	46.0	<1	92.9	1

In addition, when the number of access routers grows, the size of the multicast tree increases at a faster rate; FRM then needs to use more redundant packets than AOM to encode the forwarding information, which explains why the performance of FRM increasingly deviates from that of AOM.

The performance of IP multicast sparse mode (IP-SM) is closely related to the selection of rendezvous point (RP). In the simulation, we select node 10 as the RP for IP-SM scheme. The efficiency of IP-SM compared to IP-DM in scenarios with sparse node distribution is clearly demonstrated in Fig. 3.15. In the mean time, IP-SM has higher W_x values than AOM and FRM in the sparse cases. The reason is that the data packets are first unicast to the RP and then disseminated to access routers from there, which causes some redundant transmissions.

Link Packet Transmission. The link packet transmission, L_x, is defined as

$$L_x = T_l/C, \tag{3.8}$$

where T_l denotes the total number of data transmissions over link l within the simulation duration and C the total number of packets generated by the source. L_x represents the average number of transmissions over a given link required to multicast a single packet from the source node to all the receiver domains, which is a good indicator of traffic load due to multicast and may be exploited for admission control [3, 11].

The distribution of L_x over those backbone links traversed by multicast packets is computed in two scenarios, with the number of access routers $N=12$ and $N=20$, respectively. The results are summarized in Table 3.2 for different multicast schemes. Specifically, the total number of links involved in multicasting, the L_x values, and the corresponding percentage of links (%) are listed. In both scenarios, about 90% of links see exact 1 transmission under AOM, and rest of the links observe 2 transmissions; the bandwidth efficiency is very close to that under the IP-SM scheme. The redundant traffic in AOM is incurred by splitting the destination

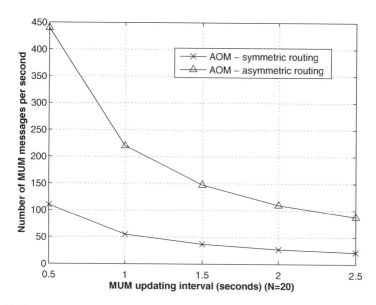

Fig. 3.16 Control message overhead

set into smaller subsets to fit into the shim header. In FRM, the largest value of L_x is up to 4 and 7, respectively, in the two scenarios, which is due to the multicast tree splitting for shim header coding.

The L_x value under IP-DM may not be integers since IP-DM periodically executes the "broadcast-and-prune" operation to establish the tree, and those pruned links only see the tree-constructing broadcast messages but no data packets. Table 3.2 shows that when the access routers become more densely populated, more links see data transmissions. Moreover, Table 3.2 shows that for the 29-link multicast tree, the "broadcast-and-prune" operation involves 50 links, which results in redundant traffic. In IP-SM, only the link between node 12 and 9 and that between node 9 and 10 observe 2 packet transmissions since every packet should be unicast to the RP at node 10 before being multicast.

Control Message Overhead. The AOM control message overhead is caused by the MUM, which is used to update membership as well as establish the routing states along the path from the access router to the source node. The control message overhead is determined by the MUM updating frequency, and extra overhead is incurred by the asymmetric inter-domain routing. As discussed in Sect. 3.4.2, for asymmetric routing, the MUM needs to travel upstream and then downstream to install IRDR_BFs on appropriate routers, and the RDR also needs to send an extra upstream message to remove the routing states constructed in the first upstream travel under the symmetric routing assumption. In the symmetric routing case, one run of upstream travel is enough.

The AOM control message overheads in both symmetric and asymmetric routing scenarios are examined in the simulation. Figure 3.16 illustrates the total number of control messages across the network per second versus the MUM updating intervals, with the number of access routers $N = 20$. We change the weights of links to realize the asymmetric routing configuration. For example, the paths connecting node $\{0, 1, 27, 28\}$ with the source node 12 are marked in Fig. 3.14. In the symmetric routing case, the MUMs go up and multicast data stream down along the links highlighted with the thick lines; in the asymmetric routing case, the downstream forwarding paths are marked with arrows. Use C_{up} to denote the number of MUM transmissions in the upstream path to the source node and C_{down} the downstream transmissions in the downstream forwarding path. According to the discussions in the previous paragraph, the ratio of control overhead in the asymmetric routing case to that in the symmetric routing case can be expressed as $\frac{2 \cdot C_{up} + C_{down}}{C_{up}} = 2 + \frac{C_{down}}{C_{up}}$, which is independent of the MUM updating interval. In our simulations generating the results of Fig. 3.16, $N = 20$, $C_{down} = 220$, and $C_{up} = 110$ and leads to a control overhead ratio of 4 determining the distance between the two curves.

By adopting the MUM updating technique suggested in [9], the size of the MUM can be controlled small as 15 bytes, leading to a total packet size of 73 bytes with TCP/IP/MAC headers taken into account. For the asymmetric routing configuration, Fig. 3.16 shows that 440 control messages are running over the network per second when the updating interval is 0.5. For a large-scale Internet backbone with 20,000 access routers, the control overhead can be estimated as $440 \times 73 \times \frac{20,000}{20} = 32.12$ MBps, which occupies only a small fraction of the bandwidth of the backbone network.

3.6.3 Forwarding False Positive

We here demonstrate the efficiency of AOM in reducing the false positives induced by the Bloom-filter scheme, with comparison to the FRM scheme.

In FRM, the Bloom filter in each packet header encodes the whole multicast tree. When a packet arrives at a TBR, all the output interfaces connecting the TBR to its neighboring TBRs will be checked to detect the tree branch. Use n_t to denote the number of tree branches encoded in the packet Bloom filter, the false positive in identifying a tree branch is $(1 - (1 - \frac{1}{m})^{n_t \cdot k})^k$. The property of FRM is that the false positives at different hops are independent, and thus the probability that an upstream false positive is propagated to downstream nodes and further to the destination is negligibly small. Therefore, the probability F_p that a given RDR receives traffic by false positive under FRM is mainly determined by the last-hop positive as $F_p \approx (1 - (1 - \frac{1}{m})^{n_t \cdot k})^k$.

We compare the F_p probability obtained under AOM with that under FRM in Table 3.3. The binary tree topology [5] with different heights is considered. The Bloom filter size in the packet header is set as 100 bytes and 20 hash functions are used. For each tree, Table 3.3 lists both the numbers of elements to be encoded and the corresponding false-positive rates.

Table 3.3 Forwarding
false-positive comparison
($m = 800$, $k = 20$)

H	AOM		FRM	
2	4	2.6934×10^{-53}	6	5.4875×10^{-50}
3	8	1.4433×10^{-15}	14	2.5713×10^{-11}
4	16	2.3242×10^{-10}	30	2.8203×10^{-6}
5	32	6.6281×10^{-6}	62	0.0085
6	64	0.0111	126	0.4172
7	128	0.4358	254	0.9658

Compared with FRM in terms of forwarding false-positive rate, AOM is more efficient since AOM packets only need to remember where to go (RDRs) instead of how to get there (the whole multicast tree) as FRM does. The efficiency of AOM over FRM will be particularly significant, when the multicast tree has dispersive branches, and each branch needs to travel a long distance in terms of hops to reach a destination domain. Such kind of tree requires a small number of RDRs to be encoded in the DST_BF and thus reduces the false-positive rate in the bit-matching operation. In contrast, as FRM is designed to encode the entire tree in each packet, it is more suitable for "short" multicast trees with fewer branches.

3.7 Summary

In this chapter, we have proposed the *AOM*, which achieves scalability by utilizing the unicast routing table and constraining the router-processing overhead. The basic idea of AOM is that if the packet carries the explicit destination addresses in its header or payload, the AON router can retrieve the destination list and leverage the unicast IP routing table to compute necessary multicast copies and next-hop interfaces. However, the fundamental issue is that we must limit the bandwidth overhead for such explicit addressing; it is impractical to attach all the destination addresses to each packet. Five fundamental practical design issues have been identified, and we have developed a Bloom-filter-based design to deal with the first two issues. With the Bloom-filter-based design, the bandwidth overhead for the explicit addressing is effectively constrained, and route aggregation and longest prefix match in the practical Internet are accommodated in AOM. The Bloom-filter-based design of AOM has the following properties: (1) The computation and memory cost incurred at each router for Bloom-filter-related process is limited and independent of the number of groups being supported. (2) The bandwidth efficiency of AOM is very close to that of the IP multicast. (3) Multicast addresses can be allocated at each source locally and decoupled with routing and forwarding.

References

1. Almeroth KC (2000) The evolution of multicast: from the mbone to interdomain multicast to internet2 deployment. IEEE Network 14:10–20
2. BGP analysis reports. http://bgp.potaroo.net/as6447/. Accessed Aug 2008
3. Bhattacharyya S (2003) An overview of source-specific multicast (ssm). IETF RFC 3569
4. Broder A, Mitzenmacher M (2004) Network applications of bloom filters: a survey. Internet Math 1:485–509
5. Chalmers R, Almeroth K (2003) On the topology of multicast trees. IEEE/ACM Trans Networking 11:153–165
6. Chu Y, Rao S, Zhang H (2000) A case for end system multicast. In: Proc. ACM SIGCOMM, 2000, pp 1–12
7. Chu Y, Rao S, Seshan S, Zhang H (2011) Enabling conferencing applications on the Internet using an overlay multicast architecture. In: Proc. ACM SIGCOMM, 2011, pp 55–67
8. Egger S, Braun T (2004) Multicast for small conferences: a scalable multicast mechanism based on IPv6. IEEE Commun Mag
9. Ermolinskiy A, Ratnasamy S, Shenker S (2006) Revisiting IP multicast. In: Proc. ACM SIGCOMM, 2006, pp 15–26
10. Fdida S, Costa L, Duarte O (2006) Incremental service deployment using the hop-by-hop multicast routing protocol. IEEE/ACM Trans Networking 14:543–556
11. Kanodia V, Cetinkaya C, Knightly EW (2001) Scalable service via egress admission control. IEEE Trans Multimedia 3:69–81
12. Kouvelas I, Fenner B, Thyagarajan A, Cain B, Deering S (2002) Internet group management protocol, version 3, Oct 2002
13. Oikonomou K, Ramakrishnan KK, Doverspike R, Li G, Wang D, IP backbone design for multimedia distribution: architecture and performance. In: Proc. IEEE INFOCOM, May 2007, pp 1523–1531
14. The network simulator – ns-2. ttp://www.isi.edu/nsnam/ns
15. Tian X, Cheng Y, Ren K, Liu B (2008) Multicast with an application-oriented networking (AON) approach. In: Proc. IEEE ICC
16. Tian X, Cheng Y, Liu B (2009) Design of a scalable multicast scheme with an application-network cross-layer approach. IEEE Trans Multimedia
17. Tian X, Cheng Y, Shen X (2010) DOM: a scalable multicast protocol for nextgeneration Internet. IEEE Network
18. Wang H, Sun T (2006) A New RP Selection Algorithm Based on Delay in DiffServ Networks. In: Proc. of IEEE International Conference on Communication Technology, Nov 2006, pp 1–4

Chapter 4
Inter-Domain AOM and Incremental Deployment

Abstract This chapter studies the rest of the practical design issues of AOM service model, including asymmetric inter-domain routing, forwarding loops prevention, and incremental deployment. In Sect. 4.1, we propose to incorporate BGP routing information to the RPF concept to address the asymmetric inter-domain routing issue, which avoids the deployment/configuration complexity of MBGP and enables a fast group joining mechanism. The false-positive property of Bloom filter can incur forwarding loops in the Bloom-filter-based multicast protocols. In Sect. 4.2.4, we analyze the effect of false positive on our most closely related work FRM. The theoretical upper bound on loops in AOM is given, and it is proven that AOM is able to automatically eliminate the forwarding loop caused by the Bloom filter false positive. Section 4.3 describes an incremental deployment solution for AOM. AOM can work over a network, in which only a small fraction of routers have AOM-aware intelligence while others are legacy routers. Extensive simulation results over a practical topology are presented in Sect. 4.4 to demonstrate the outstanding performance of AOM, with comparison to classic IP multicast and FRM. This chapter is summarized at the last section.

4.1 BGP-View-Based Joining Process[1]

The essence of the joining process described in Sect. 3.3.2 is to construct a SRC-based reverse SPT, which requires the path from the SRC to a RDR is symmetric to the one used to go from the RDR to the SRC. Nevertheless, the inter-domain routing is usually asymmetric because of administrative reasons [8]. In Fig. 4.1, it is possible that the MUM sent by E takes the path E-C-A to reach S, while the downstream data path is A-B-E. We note that an alternate option might be to use the Multiprotocol Extensions to BGP-4 (MBGP) [1, 2, 8] announcing different

---

[1] ©[2010] IEEE. Portions reprinted, with permission, from [18].

X. Tian and Y. Cheng, *Scalable Multicasting over Next-Generation Internet: Design, Analysis and Applications*, DOI 10.1007/978-1-4614-0152-0_4, © Springer Science+Business Media New York 2013

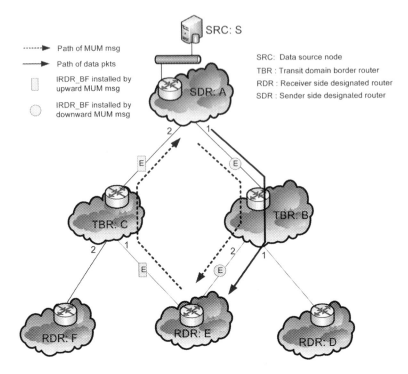

Fig. 4.1 The basic principles of AOM

unicast- and multicast-capable routes, which enables the MUM to take the correct joining path up to the SRC; however, this introduces the deployment and configuration complexity of MBGP [1, 8] which we intend to avoid.

4.1.1   Revisiting the Basic Design of AOM

The basic design of AOM proposed a round-trip join message solution to deal with the asymmetric routing challenge in Sect. 3.4.2. Figure 4.1 shows an example of AOM joining process in the asymmetric routing scenario, where the MUM takes a round trip to set up an IRDR_BF at appropriate router interfaces. The MUM message is first unicast to the SRC in the same manner described previously; however, the data packets from the SRC to the RDR may follow a path that is different from the upward-MUM-established path for some administrative reasons [8] (e.g., the link between A and C is not allowed for multicasting). In this case, besides forwarding the MUM to the SRC, the SDR further passes the MUM along the path the downstream data packets will actually take so that the links multicasting traffics to traverse are labeled with corresponding IRDR_BFs. As an extra step, the RDR then sends a special MUM to the SRC to remove the IRDR_BFs that had been established under the symmetric assumption. The up- and downward dotted lines in the right part of Fig. 4.1 demonstrate an example of this process; the IRDR_BFs represented with shadowed rectangles are to be deleted by the special MUM message.

We can see from the demonstration in Fig. 4.1 that AOM joining process is time-consuming. This is due to two reasons: first, AOM constructs the reverse SPT possibly in the asymmetric routing scenario, which contradicts the symmetric routing precondition of the reverse SPT; second, even in the symmetric routing scenario, the joining process has to be completed at the SRC node, as intermediate routers are only with destination-specific forwarding states. The following sections describe the BGP-view-based joining scheme to conquer these problems, where there are two prerequisites: (1) the physical links of the data delivery path from the SDR to an RDR must be bidirectional and (2) the inter-domain routing policy must allow control messages (e.g., MUMs) going along the path that is reverse to the data delivery path. Considering the investment efficiency for link deployment, as well as the remarkable convenience to be gained in forwarding large quantities of data packets by the inter-domain routing policy, these two conditions are quite realistic.

4.1.2   Efficient Reverse SPT Constructing

The BGP-view-based joining process could construct the reverse SPT even with asymmetric inter-domain routing, as illustrated in Fig. 4.2. The service provider designates a BGP-speaking SDR, which allows the SDR to compute the shortest paths from itself to any possible receivers. The information is stored in the local BGP routing table, where each table entry represents the local routing view for a given destination network prefix. For instance, the BGP routing entry for the network associated with E shows that E can be reached through the next-hop B and the path vector B-E in Fig. 4.2. The BGP routing entry is notified to the corresponding RDR so that the receiver side knows the actual routing view the sender-side can see. Then, the MUM is forwarded along the reverse path indicated by the BGP path vector with source routing, rather than the path indicated by the unicast routing table. In our example, the MUM takes the path E-B-A instead of E-C-A in Fig. 4.2. The MUM can correctly install forwarding states at corresponding routers along the indicated joining path to the SRC, as the dashed-line path in Fig. 4.2.

A natural question to ask is as follows: how the BGP view seen by the SDR is notified to an RDR? The key observation is that AOM adopts the *source-based* service model [4, 9], where a receiver application must know the SRC information (i.e., SRC IP address, channel number, etc.) before subscribing to a channel. A number of techniques can be used to transport the BGP routing entry from the SDR to a RDR, including via web pages, sessions announcement applications, etc. [4]. We note that there should be an interface between the SDR and multicasting

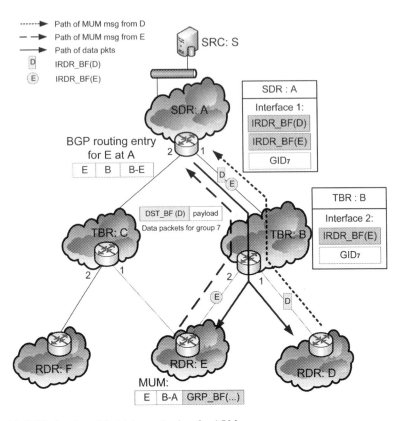

Fig. 4.2 BGP-view-based fast-join mechanism for AOM

applications at SRC to retrieve BGP views considering the BGP routing selection rules [12]; however, the implementation details are out of the scope of this thesis. We would like to emphasize that the BGP view retrieval and notification incur no extra joining delay to AOM, since the joining action starts only after the SRC information obtaining process completes, which is unavoidable for any *source-based* multicasting service model.

4.1.3   Fast Group Joining

A side benefit, derived from AOM's nature to install forwarding states along data delivery paths, is that AOM enables a fast group joining procedure. It is possible for the RDR to start receiving requested packets before its MUM message arrives at the SRC. This is achieved by temporarily labeling the explicit group IDs at the intermediate routers during the joining process.

A sample fast group joining procedure is illustrated in Fig. 4.2. After sending initial MUMs toward a given SRC, a rudimentary multicasting tree is established between the SRC and subscriber RDRs. In Fig. 4.2, a subsequent MUM from E can follow the same path marked by the dashed-line to subscribe to a newly activated group within its domain, say group 7. Knowing there has been a state IRDR_BF(E), the coming MUM only places the requested group ID, that is, GID_7, at interface 2 of B. B continues to forward the MUM up to S with the same operation conducted at each router passed.

The AOM forwarding protocol needs some modification to support the fast group joining. The regular TBR/SDR forwarding process only involves the comparison between the DST_BF and IRDR_BFs. In order to facilitate the fast joining scheme, a GID-based forwarding is added and serves as an assistant process to regular forwarding. In Fig. 4.2, let B be forwarding data packets of group 7 to D along its interface 1. B discovers that the group-7 data packets that are being forwarded via interface 1 match the GID_7 labeled at the interface 2. It will immediately forward the same data packets via the interface 2 to E, although DST_BF(D) and IRDR_BF(E) do not match. Thus, E can receive the requested data packets before the MUM arrives at S.

An analysis of the the forwarding process described above reveals that the GIDs need to be stored in the router only temporarily. If the regular forwarding process confirms that the packet with GID_i should be dispatched through the interface, the GID-based forwarding process is unnecessary, and corresponding label GID_i on the interface can be deleted. Consider the group-7 packet at SDR A in Fig. 3.8, as DST_BF(D) matches the IRDR_BF(D), the regular forwarding process confirms that the group-7 packet should be dispatched through A's interface 1, GID_7 and can be deleted immediately. In TBR B, however, the GID_7 can be deleted only after the first packet with the updated DST_BF(D,E) arrives. At that time, the regular forwarding process will confirm that group-7 packets should be forwarded via interface 2 and GID_7 is no longer useful.

GID labels are only stored at routers for a short period of time; thus, the scalability of AOM is maintained. In order to constrain the bandwidth overhead incurred by the explicit enumeration of group IDs in the MUM, the fast group joining scheme is activated within a domain only after the RDR has already subscribed to the SRC for a period of time, and there have been a reasonable number of active groups running over the network. In this case, explicit enumerating a small quantity of group IDs is more bandwidth efficient than using the GRP_BF-based method.

4.2 False-Positive Analysis[2]

The Bloom filter incurs false positive [5]: it is possible that an element not encoded in the Bloom filter can be falsely detected. The theoretical analysis in the previous

---

[2] ©[2012] IEEE. Portions reprinted, with permission, from [16].

chapter focuses on the AOM service model. This section first illustrates how the forwarding loop could happen in elaborated AOM protocol and how serious the consequence of the loop will be. We then show that design of AOM could facilitate preventing the permanent forwarding loop in all cases except a subtle conservation of bits one, based on which we present the upper bound of the probability for the permanent loop in AOM. After identifying that the root cause of the loop is false-positive forwarding, we then first describe a tree branch pruning scheme based on the RPF forwarding mechanism in AOM and show that the scheme can delete the loop caused by false-positive forwarding.

4.2.1 False-Positive Forwarding on an Interface

In the AOM forwarding process, some RDR represented by the IRDR_BF may be falsely detected in the DST_BF; thus, a packet copy will be falsely forwarded via the corresponding interface. In Sect. 3.5, we analyzed the false-positive rate of bit matching between two Bloom filters. We use $f(n_1, 1)$ to denote the false-positive rate for bit matching between two Bloom filters, with each contains n_1 and 1 element, respectively. Based on the result, the rate of false-positive forwarding on an interface for AOM is

$$F_{\text{AOM}} = 1 - (1 - f(n_1, 1))^{n_3}, \tag{4.1}$$

where n_1 is the number of RDRs contained in the DST_BF, 1 means that only one RDR is encoded into each IRDR_BF, and n_3 is the number of IRDR_BFs stored on the given interface. The term $(1 - f(n_1, 1))^{n_3}$ is the probability that none of n_3 IRDR_BFs installed at the interface incur false positive when performing bit matching with the DST_BF in the packet.

Correspondingly, comparing the FRM shim header with the Bloom filter encoding a neighbor edge of the border router also possibly incurs the false-positive. The rate of false positive forwarding on the neighbor edge for FRM is

$$F_{\text{FRM}} = f(n_2, 1), \tag{4.2}$$

where n_2 is the number of tree branches encoded in the shim header of the FRM packet. The physical meaning of (4.2) is that the rate of false-positive forwarding on the neighboring edge under FRM equals the false-positive rate of bit matching two Bloom filters, which encode n_2 and 1 element, respectively.

The Bloom filter size can be computed if the desired false-positive rate level is given [5]. For a fixed-size shim header containing a specific number of elements, the probability of false-positive occurrence is higher for bit matching with multiple single-element Bloom filter than for bit matching with only one single-element bloom filter. That is, if $n_1 = n_2$, $F_{\text{AOM}} \geq F_{\text{FRM}}$; however, covering n_1 destinations usually requires n_2 ($n_2 \geq n_1$) branches in the multicast tree [6], which could lead to

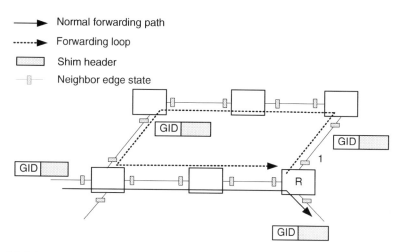

Fig. 4.3 Directed cycle caused by false positive forwarding

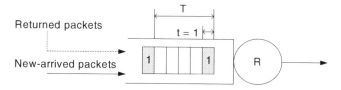

Fig. 4.4 Effect of the directed cycle on network performance

$F_{AOM} \leq F_{FRM}$. We gave the numerical analysis of a binary tree topology for both AOM and FRM in [17]. For FRM, if it wants to maintain a low false-positive rate, the shim header cannot encode too many branches; thus, more redundant traffic will be generated, as shown in Sect. 4.4.3.

4.2.2 Forwarding Loop Issue of FRM

We identify an important issue of FRM: the false-positive forwarding on the neighbor edge can possibly form the forwarding loop. We analyze the effect of the forwarding loop issue in this subsection and theoretically prove that the loop can be automatically removed in AOM in the following discussion.

As depicted in Fig. 4.3, the false-positive forwarding can happen at each interface independently; the forwarding loop is formed when the falsely forwarded packet keeps mismatching with neighbor edge states installed along the interfaces that constitute the loop topology. Suppose the network with a forwarding loop is processing only one packet, as illustrated in Fig. 4.4. We only consider the transmission delay and assume the transmission delay of every router in the network is $t = 1$ s. In our example, if the loop exists, the returned packet 1 will arrive at router R every T seconds, and $1/T$ of R's processing time is used to deal with the redundant packet 1 caused by the loop. Now, we inject other $T - 1$ packets with different group IDs to this system it is obvious that router R can only handle the returned packets from the forwarding loop. In practice, the router capacity could be large, but the injecting traffic could also be huge. The consequence of the forwarding loop can be serious: the involved routers will soon be overwhelmed by the redundant packets returned along the forwarding loop, which leads to the partial breakdown of the network.

The forwarding loop in the network normally can be eliminated by reducing the value of time-to-live (TTL) field in the IP header. However, the value of TTL field in the inter-domain routing scenario is usually large. If such loops are small, the TTL value reduction may not prevent substantial unnecessary looping over domains[5]. Moreover, the error-prone configuration of inter-domain routing can also cause the innate Internet loop prevention scheme ineffective [14]. Considering all these reasons, it is necessary to purposely design a scheme for the Bloom-filter-based multicast protocols to eliminate forwarding loops.

LIPSIN proposes to deploy a FRM-like multicast protocol in a publisher/subscriber network fabric [13]. Each link of the network is assigned d IDs. Thus, there are d candidate in-packet Bloom filters for a given multicast tree, from which a bloom filter with the best false-positive performance can be selected. When receiving a packet, the router analyzes the in-packet bloom filter to check if it contains a path that may lead the packet to return. If positive, the packet and its incoming interface will be cached. A loop is detected if the packet with the cached in-packet Bloom filter returns to the router from an interface other than the cached one. Nevertheless, the router caching the suspect packet is not necessarily the origin of the loop; therefore, the false-positive traffic cannot be fully truncated. To deal with the challenge, the caching router has to signal a request upstream toward the data source to insert a different bloom filter in the packet for the multicast tree, which imposes much burden on the data source node; moreover, there is no way to guarantee that the re-encoded Bloom filter will never incur forwarding loop at a different router in the network.

BloomCast proposes a *bit permutation* technique to reduce the Bloom filter false-positive effect. Different from FRMs directly encoding tree branches at source node, BloomCast let joining messages record each hop they traveled starting from leaves of the tree, encode the hop in a Bloom filter, and remap the Bloom filter to a different arrangement at each intermediate router. A unique reverse shortest path tree (SPT) is then created at the data source node by ORing all cumulatively permuted Bloom filters in joining messages. During the forwarding, the falsely delivered packet cannot be correctly de-mapped through the bit permutation at each hop, so the packet with no matched output interfaces will be dropped. Unfortunately, although BloomCast works smoothly under the symmetric routing assumption, where the shortest path from node A to B is the same one used to go from B to A, the inter-domain routing is usually asymmetric for the administrative reasons [8].

Further, BloomCast still can not identify the origin of the forwarding loop once it occurs, and bit permutation can only mitigate the probability of the forwarding loop rather than totally prevent it.

4.2.3   Theoretical Analysis of AOM on Loops

Automatical Loop Elimination with AOM

Lemma 4.1. *The AOM downstream forwarding can eliminate forwarding loops incurred by the false-positive forwarding under the following conditions (i) the domains associated with the SDR and RDRs are stub domains of the multicast group under consideration and (ii) the number of 1-bit positions in the DST_BF will decrease if the multicast flow diverges at some node.*

Proof. Because of condition (i), the loop can only happen among TBRs [17]. Similar to the case in Fig. 4.3, the forwarding loop in AOM is formed if some falsely forwarded packet keeps mismatching its updated DST_BF with IRDR_BFs along the interfaces that constitute the loop topology. However, as the AOM protocol incorporates the DST_BF updating operation, the number of 1-bit positions in the DST_BF will continuously decrease as the packet travels around the loop. Specifically, for a given IRDR_BF that is falsely incorporated into the updated DST_BF, there must be a TBR in the loop, which will direct at least one of the packet copy out of the loop and remove the corresponding IRDR_BF from the original DST_BF. This is because any IRDR_BFs cannot be installed to form a loop according to the forwarding states establishment process in AOM. In addition, the updated DST_BF for the packet copy that remains in the loop can only contain less 1-bit positions than the original DST_BF according to condition (ii). Consequently, the packet copy remains in the loop will have fewer and fewer 1-bit positions in its DST_BF. When the remained 1-bit positions cannot match any IRDR_BF in a TBR, the packet is dropped and the forwarding loop is eliminated.   □

The condition (ii) of Lemma 4.1 seems to be redundant, because if the multicast flow diverges, the corresponding BRA_BF along each tree branch will contain less RDRs, and the number of 1-bits representing those RDRs will be set to 0. However, when the false positive comes into play, this may not be the case. Consider the subtle case illustrated in Fig. 4.5. The packet should have gone through interface 2 of TBR B, but the false-positive forwarding occurs along interface 1; moreover, the yielded BRA_BF happens to set all 1-bit positions the same as in the incoming packet DST_BF. If the yielded data packet keeps mismatching along a loop topology and the all 1-bit positions remain unchanged, the DST_BF updating operation will not be able to eliminate the false-positive forwarding loop. This kind of event is termed as *conservation of bits*.

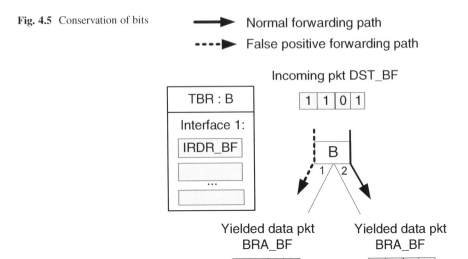

Fig. 4.5 Conservation of bits

Upper Bound on Loops in AOM

Theorem 4.1. *The upper bound of the probability that a permanent loop occurs in AOM is*

$$\left(\sum_{i=1}^{M} \binom{M}{i} \cdot P_f^{\,i} \cdot (1 - P_f)^{M-i} \cdot \left(1 - \left(1 - \frac{k}{m} \right)^i \right)^X \right)^3 \tag{4.3}$$

under the conditions (i) in Lemma 4.1, where

- *M is the maximum number of IRDR_BFs on an interface of a TBR in the network.*
- *P_f is the false-positive rate for bit matching between two Bloom filters, one containing the maximum number of elements in DST_BF and the other 1 element.*
- *X is the maximum number of 1-bit positions in a DST_BF.*
- *m is the total number of bit positions in a DST_BF or IRDR_BF.*
- *k is the number of hash functions to form a DST_BF or IRDR_BF.*

Proof. According to previous lemma, the permanent forwarding loop occurs only if the conservation of bits event illustrated in Fig. 4.5 happens along a loop topology; because the DST_BF updating operation of AOM can always eliminate forwarding loops in other cases automatically, the probability for conservation of bits event happens along a 3-node loop topology must be the upper bound for a permanent loop occurs.

We first find the probability that the conservation of bits event happens on one node. Specifically, suppose we have a DST_BF with X bits of 1s, and there are M IRDR_BFs on the local interface. We want to find the probability that at least one of the IRDR_BFs along the interface incurs false-positive forwarding and those falsely matched IRDR_BFs form a BRA_BF that is the same as the original DST_BF. This probability is

$$\sum_{i=1}^{M} \binom{M}{i} \cdot P_f^{i} \cdot (1 - P_f)^{M-i} \cdot P(e),\qquad(4.4)$$

where e is the event: given a set of i falsely matched IRDR_BFs, a DST_BF has at least one of the IRDR_BF match all bit positions in itself, and $P(e)$ is the probability that event e happens.

$P(e)$ can be found through $1 - P(\bar{e})$. According to the principle of the Bloom filter, an IRDR_BF has probability $\frac{k}{m}$ to set a single bit position the same as in the original DST_BF; thus, the probability that given i falsely matched IRDR_BFs, none of the IRDR_BF matches *a single 1-bit position* in the DST_BF is $(1 - \frac{k}{m})^i$.

For X bit positions in the original DST_BF, the probability that each 1-bit has at least one of the i mismatched IRDR_BFs setting the same 1-bit position is

$$P(e) = \left(1 - \left(1 - \frac{k}{m}\right)^i\right)^X.\qquad(4.5)$$

Consequently, the probability that the conservation of bits event happens on one node is

$$\sum_{i=1}^{M} \binom{M}{i} \cdot P_f^{i} \cdot (1 - P_f)^{M-i} \cdot \left(1 - \left(1 - \frac{k}{m}\right)^i\right)^X.\qquad(4.6)$$

Thus, the conservation of bits event over 3-node loop happens with the probability, as shown in (4.2).

The upper bound on permanent loops in AOM can be very low. Suppose we use the same Bloom filter configuration as in FRM [7], and the network topology adopted in [16], the resulted upper bound is in the order of 10^{-36}.　□

4.2.4 Complete Loop Elimination in AOM

Pruning False Tree Branches. In AOM, a RDR stored on an interface of a router (in the form of an IRDR_BF) implies that a forwarding path through the interface exists from the router to the destination domain represented by the RDR, according to the joining process and the RPF techniques adopted in AOM. Note that such a fact is true in both symmetric and asymmetric scenarios [16]. Thus, if a false-positive forwarding happens in the network, due to mismatching with a certain RDR$_i$ on

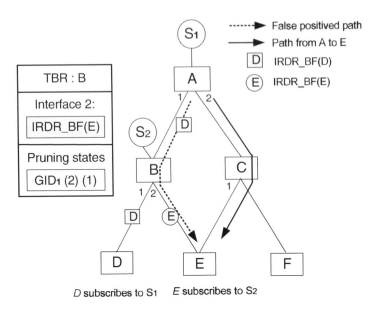

Fig. 4.6 Pruning false tree branches

an interface, the falsely forwarded packet will finally reach the destination domain associated with RDR_i. The destination domain can then identify that the traffic was due to false forwarding if the group was not requested by it, and subsequently sends a pruning message upstream reverse to the forwarding path to prune the false-forwarding branches and stop the misdelivered packets.

Implementing the branch pruning design is not trivial. How to ensure the direction of the upstream pruning messages along those false-positive branches? Consider the example shown in Fig. 4.6, where RDR D and E subscribe to SRC S_1 and S_2 along path D-B-A and E-B, respectively. The packet generated by S_1 may be falsely forwarded to E if a false-positive match with IRDR_BF(E) on interface 2 of B happens. When E received the misdelivered packet associated with S_1, it cannot tell that the false forwarding was induced by a joining message to which SRC (note that E may have sent joining messages to many different SRCs). Even if E somehow correctly sets S_1 as the destination of the pruning messages, there are multiple paths from E to S_1. If E let pruning messages to S_1 take E-C rather than E-B as the reverse path, the pruning operation still cannot block the misdelivered traffic actually along B-E. With limited knowledge, the destination SRC of the upstream pruning message could not be properly set, and the upstream path could not be determined with the impact of asymmetric routing.

We design the pruning message propagation scheme as follows. In the first hop, the destination domain receiving falsely forwarded packets just sends a pruning message upstream through the interface where the false packets come. The pruning message carries the GID associated with falsely forwarded group. When the pruning message reaches an intermediate router, the false packets will be still coming to the router. By checking the incoming interface for packets associated with the tagged GID, the intermediate router can then easily identify the next upstream hop. Note that the destination domain receiving false packets will keep sending pruning message upstream, according to a certain schedule such as sending one pruning message after receiving n (≥ 1) false packets, until the false packets stop coming.

To facilitate the pruning process, each router involved also manages pruning states. When the pruning message arrives at an intermediate router for the first time, the router creates a pruning state in the format of $\text{GID}_i(\mathscr{F})(\mathscr{N})$, where GID_i is the group identifier of the misdelivered packets, $\mathscr{F}$ is the set of output interfaces that have falsely forwarded packets with GID_i, and $\mathscr{N}$ is the set of output interfaces that are normally forwarding packets with GID_i. If there are subsequent pruning messages associated with GID_i coming from another interface, the corresponding interface will be moved from set $\mathscr{N}$ to the set $\mathscr{F}$. In the example shown in Fig. 4.6, $\mathscr{N} = 1$ and $\mathscr{F} = 2$, which means that the interface 1 of B is normally forwarding packets of GID_1, because otherwise there should be another pruning message from interface 1. In this way, B could block the traffic with GID_1 toward E and keep forwarding that toward D.

Operations on Pruning States. The pruning states can help a TBR to properly determine whether to continue forwarding the pruning messages upstream, stop the forwarding, or remove the obsolete pruning states. Specifically, a TBR will take the following options upon receiving a pruning message associated with GID_i:

- Continue forwarding upstream the pruning message, if $\mathscr{F} \neq \phi$ and $\mathscr{N} = \phi$. This condition indicates that the TBR is an intermediate router along the false-forwarding branch, so it needs to continue forwarding the pruning message upstream toward the root node that generates the false-forwarding traffic.
- Stop forwarding upstream the pruning message, if $\mathscr{F} \neq \phi$ and $\mathscr{N} \neq \phi$ or the TBR is an SDR of a source domain. The first part of the condition, that is, $\mathscr{F} \neq \phi$ and $\mathscr{N} \neq \phi$, indicates the current TBR is the root node that generates the false-positive match, so there is no need to further forward the pruning message upstream. The branches in set $\mathscr{N}$ indicates the normal paths that correctly forward traffic for group GID_i, while branches in $\mathscr{F}$ indicates paths due to false-positive match. Such a situation could only be possible that the current router receives correct traffic but false-positive matching happens in the forwarding stage, that is, the current TBR is the root for false-positive traffic. The second part of the condition is due to the possibility that the pruning message may go up to the SDR in the source domain if the false positive happened due to the mismatch when checking GRP_BF against the local channel list.

It is not difficult to see that the pruning state associated with a GID on an interface should be removed, if the destination domain ending the false-forwarding path now actively requests traffic from this group. Thus, when a destination domain needs to join a new group, it first check whether the group was involved in the false-positive situation before. We let the RDR keep a record of the false-positive GIDs it has observed. If the new active GID is found in the record, it needs to use a joining message carrying explicit GID and a pruning-removing flag to remove the pruning states on related interfaces. The destination address of such a joining message is set as the SRC address associated with the GID and then follows the AOM joining procedure (note that BGP-view-based joining is applied in the asymmetric case [16]). When a TBR/SDR receives a joining message with a pruning-removing flag, if it has a pruning state associated with the indicated GID, it then just remove it. Note that such a pruning removing procedure is efficient, which just remove the pruning states on related hops that might impact the normal forwarding. Considering the false-positive probability is small, the joining messages and computing overheads in intermediate TBRs associated with these falsely delivered groups will not impact the scalability much.

Complete Loop Elimination

Theorem 4.2. *The false tree branch pruning scheme can stop the false traffic and eliminate forwarding loops incurred by the false positive in Bloom filter matching under the condition that the domains associated with the SDR and RDRs are stub domains of the multicast group.*

Proof. Note that any IRDR_BF on a certain interface was placed by a joining message from a destination domain; reverse to the path of the joining message is a data forwarding path to the destination domain according to the AOM design for both symmetric and asymmetric cases [16]. Therefore, for given IRDR_BF that is falsely incorporated into the updated DST_BF, there must exist a TBR in the loop, which has a branch leading to the destination domain that generated the join message associated with the IRDR_BF under consideration. So the redundant traffic due to false positive can definitely be detected by that destination domain and incurs subsequent pruning messages. According to the pruning mechanism design, the pruning message will finally reach the root node that originated the false-positive traffic and stop the traffic. Thus, all the false-positive traffic downstream and the forwarding loop if any will be totally eliminated.   □

Consider the example shown in Fig. 4.3, the false-positive matching is initiated at the router R, and a forwarding loop is further formed, as shown in the dashed line. In this example, the router R' has a branch inducting the false-positive traffic to a destination domain. The destination domain will then identify the situation of false positive and generates the pruning messages. The pruning messages will reach the root node R of the false-positive traffic and stop the loop.

4.3   Incremental Deployability

The AOM protocol we presented in the previous chapter requires every router in the network to be aware of AOM. This introduces a deployment problem as it is impractical to update the long existing legacy infrastructures simultaneously. We propose to use IP tunneling technique to establish a AOM-aware tunnel system across the entire network regions, where AOM-aware routers and legacy routers coexist. The basic idea is to consider the unicast path connecting two AOM-aware routers as a *logical interface*. When receiving a AOM packet, a AOM-aware router can conduct the same aggregation, replication, and MGL record updating operations but based on the states installed for the logical interfaces. The generated packets copies are then dispatched along the logical interface that leads to the next-hop AOM-ware router. Routers that do not implement AOM protocol will forward the packet as if they are doing regular unicast.

Figure 4.7 illustrates how to incrementally deploy AOM. The black rectangles in the figure represent legacy unicast routers that do not implement AOM. When RDR G and H want to subscribe to the SRC S, they send out the MUMs toward S. A little modification required is to allocate a *logical interface* field in the packet and let the MUM record the local prefix each time it is delivered by the AOM-aware router. For example, in Fig. 4.7a, when the MUMs are sent from G and H, as they are AOM-aware routers, the prefixes of G and H are recorded as the logical interfaces in the two MUMs, respectively. After the MUMs arrive at B, B discovers that although the two MUMs come from the same physical interface 1, they are actually from two different logical interfaces. B creates two states, that is, IRDR_BF(G) and IRDR_BF(H), for these two logical interfaces. Then, B continues to forward the MUMs up to A. The logical interface field of both MUMs is set to B at this time, since B is also aware of AOM. When the MUMs reach A, A finds out that the two MUMs are from the same logical interface B, and corresponding states are generated, as shown in Fig. 4.7a. In this way, each intermediate AOM-aware router knows the next-hop AOM-aware router and the AOM-aware tunnel system is established, which will facilitate the multicast data forwarding.

Figure 4.7b depicts the forwarding procedure of the incremental deployment solution. When A receives the multicast packet with destinations G and H, it checks the forwarding states installed for each of its logical interfaces as regular AOM forwarding process. A notices that G and H could be reached through the same logical interface B, so only one packet is forwarded through the logical interface B. This packet is firstly encapsulated into an ordinary unicast packet with the logical interface B as the destination (recall that B is actually the network prefix of the transit domain border router), and then unicast to B, as shown in Fig. 4.7b. After this packet arrives at B, as B is AOM aware, it decapsulates the multicast packet and performs the same forwarding operations, as described earlier. Two packet replicas are to be generated according to the states installed for the two logical interfaces at B, and these replicas are encapsulated and delivered to corresponding next-hop AOM-aware routers. These packets will pass a legacy router in the following forwarding process. The legacy router is unable to recognize the AOM-type packet and does not decapsulate the received packets; it just forwards the packets based on the packets' destination addresses.

An interesting property of the incremental deployment solution is that the number of forwarding states installed at a given AOM-aware router only depends on the number of RDRs which could be reached through this router, and independent of the fraction of AOM-aware routers within the network. As shown in Fig. 4.7b, there will be two forwarding states for RDR G and H, no matter the intermediate router toward B is aware of AOM or not.

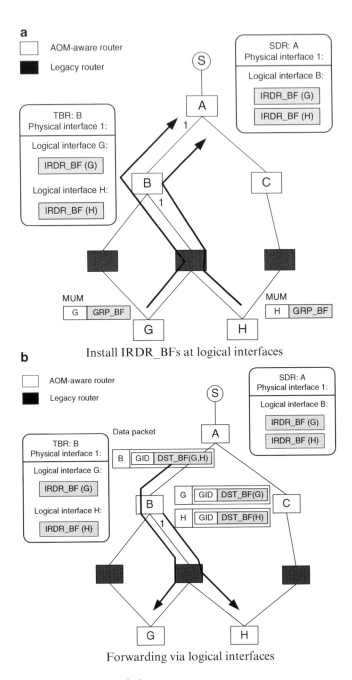

Fig. 4.7 Incremental deployment solution

With the logical interface, based approach, the fast group joining scheme described earlier is also workable, and the only variation is that the GID labels are operated according to the logical interfaces. In the incremental deployment solution, AOM-aware routers generate packet replicas only when it is necessary. The legacy routers do not affect the network regions where AOM-aware routers are concentrated, but they could lead to more redundant traffic (e.g., the multi-unicast at *B*) without affecting the correctness of AOM protocol.

4.4 Performance Evaluation

4.4.1 Simulation Setup

We use ns-2 [15] simulation results to demonstrate the performance of BGP-view-integrated AOM in this section. The network topology for simulation is given in Fig. 4.8, which extends the one used in the previous chapter. In this model, the source and transit domains are represented as backbone routers or backbone nodes; similarly, regional and organization autonomous systems (ASes) are represented as designated border routers in those domains. The backbone router providing connection service to regional ASes is termed as *access router* or *access node*, and regional ASes are connected by organization ASes. Nodes 2 and 11 in Fig. 4.8 are two examples to illustrate the connection relationship. We simulate multisource multigroup scenarios where SRCs are located at nodes 12 and 24

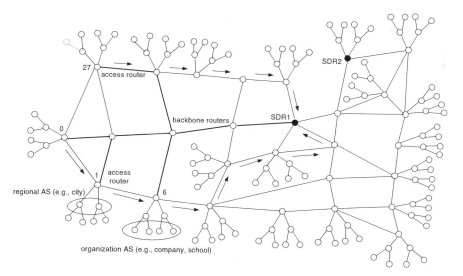

Fig. 4.8 Simulation topology

with each providing 15 groups and RDRs are placed at organization ASes. In the simulation, each RDR randomly chooses 10 groups at each SRC to subscribe to. After the rudimentary multicasting tree is established, RDRs send subsequent MUM messages to join in the other 5 groups provided by each source. To study the scalability of the proposed scheme, we incrementally add subscribing RDRs from different areas of the topology. We analyze the average delay experienced by all of the RDRs over all of the groups using AOM with and without the fast-join mechanism; we also analyze the average delay distribution among all participating RDRs.

4.4.2   Memory Overhead

To compare the memory overhead of AOM with other multicast mechanisms, we count *the number of forwarding entries at backbone routers and regional AS border routers that are involved in the multicast*. The cumulative distribution function (CDF) of the number of forwarding entries per node for the multicast schemes under study is illustrated in Fig. 4.9. Scenarios that RDRs are sparsely ($N = 16$) and densely ($N = 80$) populated are considered, where N is the number of subscribing RDRs.

The forwarding state of AOM is made up of the IRDR_BF, and hence, the number of forwarding entries per node is the number of RDRs whose joining paths pass through this node. The maximum number of states at a node is equal to the maximum number of subscribing RDRs; therefore, all nodes have entries no greater than 16 and 80 in Fig. 4.9a, b, respectively. The forwarding state of FRM [7] is in fact its AS neighbor edges, and hence, the number of forwarding entries per node is the AS degree of this node. The distribution of the number of forwarding entries per node is fixed for FRM as long as the topology is given. There is no node which has more than 5 forwarding entries for FRM. IP multicast dense mode (IP-DM) and sparse mode (IP-SM) also have fixed number of forwarding entries per node, because the number is dependent on the number of groups active ($G = 200$) in the network.

Figure 4.9 shows that AOM can significantly reduce the number of forwarding states stored at each node compared with IP multicast. This is because AOM stores only one state on each related node for each subscribing domain. In contrast, each subscribing domain may join in tens of thousands of groups, and each group needs a state on each related node under IP multicast. The number of forwarding states per node for AOM is independent of the number of groups being supported by the node. AOM requires comparatively more forwarding states than FRM does; however, the states maintained can greatly benefit the bandwidth efficiency, which is to be discussed in Sect. 4.4.3.

We use back-of-the-envelope calculation to show that the memory cost of AOM is tractable. Suppose we use a 100-byte DST_BF within each packet to carry the Bloom filter for multicasting as in [7]. As AOM implementation incurs Bloom filter

Fig. 4.9 CDF of the number
of forwarding entries per
node

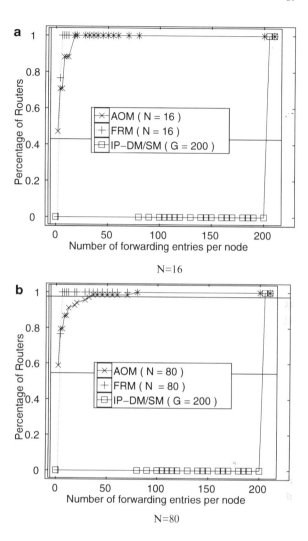

match operation, the local IRDR_BF should also be 100 bytes. The real BGP table
[3] suggests that the number of RDRs can be $O(10^5)$. Then, the IRDR_BF memory
overhead at each interface will be $O(100 \times 10^5) = O(10M)$ bytes, which can be
easily accommodated by a line card [7].

4.4.3   Bandwidth Overhead

We examine the bandwidth overhead incurred by data forwarding in this section.
The AOM forwarding incurs bandwidth overhead due to two reasons: (1) each

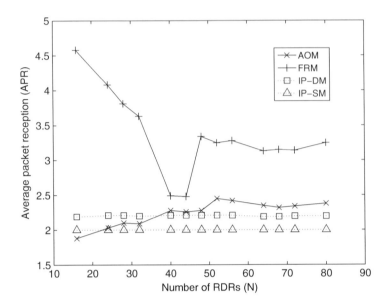

Fig. 4.10 Average packet reception (APR)

packet needs to carry a DST_BF as destinations information and (2) when the
number of destination domains is too large, multiple DST_BFs have to be used
to cover all the destinations, which will generate redundant traffic. The similar
approach has been used by FRM to use multiple packets to carry a big source-
routing tree. We assume fixed 40-byte shim header (the structure to contain
forwarding information, e.g., the DST_BF for AOM and the Tree_BF for FRM) and
measure the overhead in the packets reception. Compared with simple enumeration
of IP addresses, DST_BF can save the packet size by about 40% and incur the false-
positive rate at the level of 10^{-4} in the simulation. FRM is one of the main reference
objects in the evaluation; we note that the FRM can be optimized in bandwidth
overhead [7]; however, these optimizations can also be applied to AOM with minor
modification. For demonstration convenience, we only consider the standard design
for both AOM and FRM.

Average Packet Reception. The average packet reception (APR) is defined as *the
average number of packets received by each non-RDR router in the multicast
tree to disseminate a single packet from each source to all receivers.* Figure 4.10
shows the APR versus N (the number of RDRs involved in multicasting) with
different multicast schemes compared. In the ideal case, each router except two
SDRs should receive two packets when delivering a packet from each SDR to all
RDRs. We observe that AOM achieves the performance close to IP multicast, and
even outperforms IP multicast when N is very small, as the IP multicast has the
tree maintenance overhead. AOM has a better performance than FRM, especially

when the multicast subtree along one interface of the SDR has many branches. This is because, for FRM, more multicast packets have to be generated to carry subtree branches so that all destination RDRs can be covered.

In contrast to IP-DM, AOM does not use the "broadcast-and-prune" [1] method to maintain the tree structure. In the sparse scenario with small N, the larger value of APR for IP-DM is due to packet transmissions during the "broadcast-and-prune" operation. The performance of IP-SM is closely related to the selection of rendezvous point (RP). In the simulation, we select the backbone router at the geometry center of the topology as the RP for IP-SM scheme. As the data packets are first unicast to the RP and then disseminated to RDRs from the RP, every router under study receives two packets including the two SDRs. For AOM in the case of $N = 16$, two SDRs receive only one packet from each other, as the destinations along each interface can be encoded into a single DST_BF. This is why AOM can perform even better than IP multicast when $N = 16$.

AOM strikes a balance between bandwidth efficiency and small Bloom filter false-positive rate, in comparison with the most closely related work, FRM. The fundamental difference between AOM and FRM is that the AOM encodes only destination prefixes in the packet shim header while the FRM encodes multicasting tree branches. The Bloom filter incurs false positive, which means the element not encoded into the Bloom filter might be falsely detected. For a fixed-length bloom filter, the more elements are encoded, the higher the false-positive rate can be. In AOM and FRM, when the number of receivers/branches exceeds the capacity of a single shim header, multiple packets are sent to cover all destinations, which are counted as redundant traffic. Since covering the same number of destinations normally requires more branches, AOM can generate less redundant traffic than FRM does if they keep the same false-positive rate. We observe that the APR of FRM is decreasing when N is between 16 and 50. This is because the RDRs in these cases are concentrated on subtrees sourced from different neighbor edges of the SDRs. The FRM shim header happened to be able to contain all branches of each subtree and the number of nodes receiving exact 2 packets increases. However, when N keeps increasing, the number of branches in the subtrees go beyond the capacity of the shim header again, and the number of redundant packets increases. Thus, the FRM curve presents the shape of a funnel. In the simulation, we use the bin packing algorithm [7, 19] to compute multiple shim headers for both AOM and FRM.

Per-Node Packet Reception Distribution. The per-node packet reception (PPR) is *the number of packets received at a given node when multicasting a single packet from each SRC to all receivers.* Figure 4.11a, b plot the CDF of the PPR for different multicast mechanisms when $N = 16$ and $N = 80$, respectively. In AOM, about 80% of the nodes receive exact 2 packets in both scenarios. There is no node receiving more than 3 packets in the case of $N = 16$ and 6 packets in the case of $N = 80$. The redundant traffic in AOM is incurred by splitting the destination set into smaller

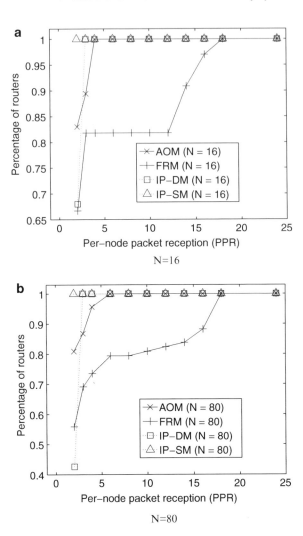

Fig. 4.11 CDF of per-node packet reception (PPR)

subsets to fit into the DST_BF. In FRM, only about 66% and 55% of the nodes receive exact 2 packets in the two scenarios, respectively. The largest number of packets a node can receive is up to 18. In IP-DM and IP-SM, almost every node receives exact 2 packets, except for a few that receive some redundant packets in the tree maintenance or the source-to-RP unicast. The number of nodes receiving no redundant packets under AOM is close to that under IP multicast and is larger than that under FRM.

4.4.4 *Joining Delay*

Average Access Delay. The average access delay (AAD) is defined as

$$\text{AAD} = \frac{\sum_{i=1}^{N} \sum_{j=1}^{M} d_{ij}}{M \cdot N}, \tag{4.7}$$

where d_{ij} represents RDR_i's access delay, that is, the delay between the MUM is sent out and the first requested data packet arrives, for group j. M is the number of groups RDR_i subscribes to, while N is the total number of RDRs in the experiment. There are two cases for RDR_i to subscribe to group j. Case i represents the scenario that RDR_i sends the initial MUM to join in group j, and case ii means that RDR_i joins in group j after the rudimentary multicasting tree is established, where the fast group joining scheme can be used. Thus, d_{ij} can be further expressed as

$$d_{ij} = \begin{cases} d_{ij}^1, & \text{if case i,} \\ d_{ij}^2, & \text{if case ii.} \end{cases} \tag{4.8}$$

We can see that $d_{ij}^1 = d_{ij}^2$ for FRM and $d_{ij}^1 \geq d_{ij}^2$ for AOM. The fundamental reason is that AOM enables intermediate routers already on the multicasting tree to respond to the joining MUM as soon as the MUM reaches it, and FRM joining scheme responds to the joining MUM only at the SRC. Since it is possible that the MUM does not meet any intermediate router already on the multicasting tree until it arrives at the SDR, d_{ij}^1 may equal d_{ij}^2 in AOM. The AAD reflects the overall timing performance of the multicast scheme, which is important to real-time applications.

In the simulation, each RDR randomly chooses 66 groups at each SRC to subscribe to. After the rudimentary multicasting tree is established, RDRs send subsequent MUM messages to join in the other 34 groups provided by each source. Figure 4.12 illustrates the resulted AADs for different quantities of RDRs. We change the weights of links to realize the asymmetric routing configuration. For example, the regular unicast paths connecting node $\{0, 1, 6, 27\}$ with the SDR1 are marked in Fig. 4.8 with arrows. As in Fig. 4.12, the use of the fast group joining scheme in AOM can shorten the AADs by approximately 22% on average, compared with the FRM.

We can see AADs of the AOM and FRM vary with the number of participating RDRs (N) non-monotonically. This is because the access delay performance is also closely related to the distance between RDRs and the SRCs. It is intelligible that d_{ij}^1 is large when RDR_i is far from the SRCs and it is small when RDR_i is near to the SRCs. With more participating RDRs near to (far from) SRCs, the AAD is reasonably shorter (longer) as the delay caused by rudimentary tree construction is shorter (longer). As an example in Fig. 4.12, for the scenario involving 16 RDRs, most of these RDRs are 5 hops away from SDR1 and 6 hops away from SDR2, and for the scenario involving 32 RDRs, some of the newly-added RDRs are nearer to SRCs, about 4–5 hops away; therefore, the AAD in the latter case is comparatively lower.

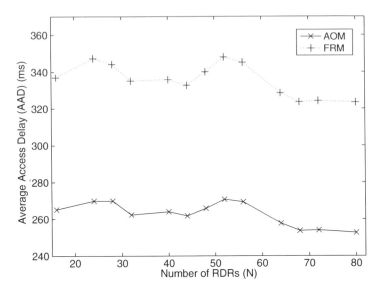

Fig. 4.12 Average access delay

Per-Node Average Access Delay (PAAD) Distribution. We here demonstrate the CDF for *the per-node access delay averaging the participated groups.* The results are depicted in Fig. 4.13. In each scenario, the curve representing the AOM is above the corresponding one representing FRM, which indicates that the number of RDRs experiencing the shorter PAAD in AOM is more than that in FRM. These results corroborate the experiment outcome demonstrated in Fig. 4.12. In sparse case of AOM, half of the participating RDRs have the PAAD that is less than 280 ms and the upper bound of the PAAD is 300 ms, as indicated in Fig. 4.13a. With FRM, RDRs having the PAAD no greater than 260 ms account for only 25% of the participating RDRs, and the PAAD upper bound increases up to 380 ms. The similar thing can also be observed in the dense case. We can see from Fig. 4.13 that the PAAD in the dense scenario is more various compared with that in the sparse scenario. This is because there are more RDRs located in different parts of the topology in the dense case, and their individual PAADs are more various in values.

4.4.5 AOM Scalability Under False Positives[3]

We now evaluate the scalability of AOM with comparison to other multicast schemes under false positives. We connect each regional AS node with 16 RDRs and let all 16 RDRs subscribe to all groups provisioned by both SRC1 and SRC2.

Fig. 4.13 CDF of per-node average access delay (PAAD)

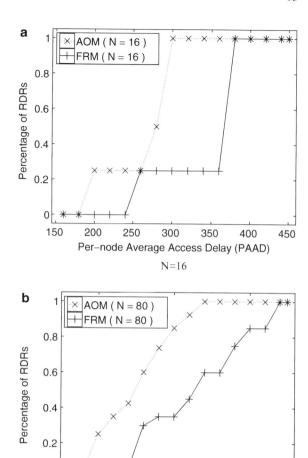

In order to clearly present the scalability of AOM under false positives, we let each SRC providing 500 groups. Then we study the impact of false-positive forwarding on memory overhead of AOM. In the simulation, the false-positive forwarding scenario is configured as follows: We let all 16 RDRs subscribe to SRC2 and only 2 of them to SRC1. This setting could establish many IRDR_BFs along links in the backbone network, which increases the probability that the SRC1 data packets are falsely forwarded to SRC2 subscribers. To make the scenario more extreme, we configure the false-positive rate for Bloom filter matching to around 20% according to (4.1) in Sect. 4.2.1. The memory overhead is also measured by counting the number of forwarding entries at backbone routers and regional AS border routers that are involved in the multicast.

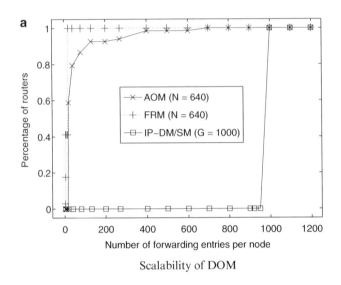

Scalability of DOM

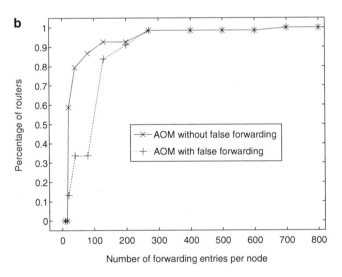

DOM with and without false positive forwarding

Fig. 4.14 CDF of the number of forwarding entries per node

The CDF of the number of forwarding entries per node for the multicast schemes under study are illustrated in Fig. 4.14a. The forwarding state of AOM with false-positive forwarding is made up of the IRDR_BF; thus, the number of forwarding entries per node is the number of RDRs whose joining paths pass through this node. The maximum number of states at a node is equal to the maximum number of subscribing RDRs; therefore, all nodes have entries no greater than 640 in Fig. 4.14. The forwarding state of FRM [7] is its AS neighbor edges, and hence, the number of forwarding entries per node is the AS degree of this node. There is no node which has more than 20 forwarding entries for FRM. IP multicast dense mode (IP-DM) and sparse mode (IP-SM) have fixed number of forwarding entries per node, because the number depends on the number of groups active ($G = 1,000$) in the network.

Figure 4.14a shows that AOM can significantly reduce the number of forwarding states stored at each node compared with IP multicast. This is because AOM states are destination specific, instead of group specific. Consequently, the number of forwarding states per node for AOM is independent of the number of groups being supported by the node. AOM requires comparatively more forwarding states than FRM does; however, the states maintained can greatly benefit the bandwidth efficiency, which is to be discussed in Sect. 4.4.3.

The impact of false-positive forwarding on AOM is illustrated in Fig. 4.14b. For the DOM with false-positive forwarding, the states $GID_i(\mathscr{F})(\mathscr{N})$ for pruning false tree branches should also be counted. Figure 4.14b shows that the pruning states have limited influence on the scalability of AOM. This is because the pruning states will be finally installed on the root node along the false-forwarding path. In this experiment, 14 of the RDRs under a regional AS are candidates for receiving unrequested data from SRC1, but there are still 2 RDRs that are normal subscribers. This means that the pruning states are primarily installed at each regional AS node. The pruning states has not been propagated into the core network, as the membership status of subscribers is not changed. We create this setting because this is an adverse case in the distribution of the pruning states. If the scalability of AOM is not serious impacted in this case, the performance for false tree branch pruning will be better in other scenarios. Imagine that if the pruning states are moving toward the backbone network, some of them will merge at some upstream node according to the pruning states operation. Although the pruning states are related to the number of falsely supported groups, the number of resulted states is still much less than that in IP multicast. The scalability of AOM remains with the false tree branch pruning operations.

4.4.6 Incremental Deployability

We here illustrate how AOM performs when only a fraction of routers are aware of AOM protocol in the entire network. We still use the topology in Fig. 4.8 and vary the percentage of AOM-aware routers from 20% to 100%. There are 2 SRCs and 80 RDRs in the simulation, where RDRs are aware of AOM and the intermediate routers are randomly assigned as AOM-aware routers for each percentage.

Effect on Memory Overhead. The incremental deployment solution does not affect the memory overhead at any given AOM-aware router. This is because the number of forwarding entries at each AOM-aware router only depends on the number of RDRs which could be reached through this router. The variation of the percentage of AOM-aware routers does not affect the number of forwarding entries at any given AOM-aware node as long as the number of RDRs and their locations are fixed.

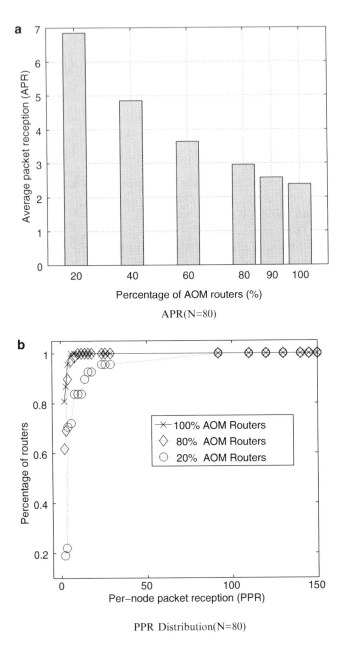

Fig. 4.15 Incremental deployment: effect on bandwidth overhead

Effect on Bandwidth Overhead. The APR and PPR distribution at different percentages are shown in Fig. 4.15, where the packet receptions of all routers are counted in. In Fig. 4.15a, the more routers are aware of AOM, the lower the APR is. This

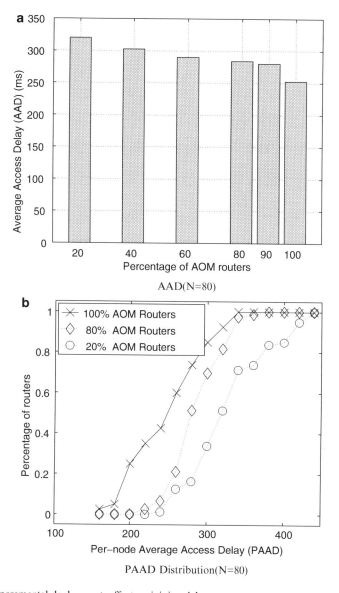

Fig. 4.16 Incremental deployment: effect on joining delay

is simply because more AOM-aware routers could reduce the chances for initiating multi-unicast over the network. In Fig. 4.15b, it can be seen that the PPR distribution in the 80% case is very close to that in the 100% case. In the 20% case, some nodes receive 30 packets when multicasting a single packet from each SRC to all RDRs, but the correctness of the protocol is not affected.

Effect on Joining Delay. The incremental deployment solution will not affect the correctness of the fast group joining scheme. Figure 4.16a illustrates the AAD with different fractions of AOM-aware routers. Even in the 20% case, it is still possible that the joining request are processed before it reaches the SRC; therefore, the AAD of AOM is no greater than that of FRM. The more routers become aware of AOM, the more likely the joining request is processed at the intermediate routers before arriving at the SRC; thus, the AAD decreases as the percentage of AOM-aware router increases. For the same reason, the number of RDRs perceived generally lower joining delay when the percentage of the AOM-aware router is higher, as depicted in Fig. 4.16b. Therefore, the fast group joining scheme works correctly even if most of the intermediate routers cannot recognize AOM protocol.

4.5 Summary

This chapter has proposed solutions to the rest of AOM practical design issues, including asymmetric inter-domain routing, forwarding loops prevention, and incremental deployment. We proposed to incorporate BGP routing information to the RPF concept to address the asymmetric inter-domain routing issue, which avoids the deployment/configuration complexity of MBGP and enables a fast group joining mechanism. The false-positive property of Bloom filter can incur forwarding loops in the Bloom-filter-based multicast protocols. We have analyzed the effect of false positive in the most closely related work FRM. Moreover, AOM has been theoretically proven to be able to automatically eliminate the forwarding loop caused by the Bloom filter false positive. An incremental deployment solution for AOM has been described in this chapter. AOM can work over a network, in which only a small fraction of routers have AOM-aware intelligence while others are legacy routers. Extensive simulation results over a practical topology have been presented to demonstrate the outstanding performance of AOM, with comparison to classic IP multicast and FRM.

References

1. Almeroth KC (2000) The evolution of multicast: from the mbone to interdomain multicast to internet2 deployment. IEEENetwork 14:10–20
2. Bates T (2000) Multiprotocol extensions for BGP-4. http://en.wikipedia.org/wiki/Bin_packing_problem. Accessed Jun 2000
3. BGP analysis reports. http://bgp.potaroo.net/as6447/. Accessed Aug 2008
4. Bhattacharyya S (2003) An overview of source-specific multicast (ssm). IETF RFC 3569
5. Broder A, Mitzenmacher M (2004) Network applications of bloom filters: a survey. Internet Math 1:485–509
6. Chalmers R, Almeroth K (2003) On the topology of multicast trees. IEEE/ACM Trans Networking 11:153–165
7. Ermolinskiy A, Ratnasamy S, Shenker S (2006) Revisiting IP multicast. In: Proc. ACM SIGCOMM, 2006, pp 15–26

8. Fdida S, Costa L, Duarte O (2006) Incremental service deployment using the hop-by-hop multicast routing protocol. IEEE/ACM Trans Networking 14:543–556
9. Holbrook HW, Cheriton DR (1999) IP multicast channels: express support for single-source multicast applications. In: Proc. ACM SIGCOMM, Aug 1999, pp 65–78
10. Kouvelas I, Fenner B, Thyagarajan A, Cain B, Deering S (2002) Internet group management protocol, version 3, Oct 2002
11. Oikonomou K, Ramakrishnan KK, Doverspike R, Li G, Wang D (2007) IP backbone design for multimedia distribution: architecture and performance. In: Proc. IEEE INFOCOM, May 2007, pp 1523–1531
12. Rekhter Y, Li T (1995) A border gateway protocol 4 (BGP-4). IETF RFC 1771
13. Rothenberg CE, Arianfar S, Jokela P, Zahemszky A, Nikander P (2009) LIPSIN: line speed publish/subscribe inter-networking. In: Proc. ACM SIGCOMM, 2009, pp 195–205
14. Särelä M, Rothenberg CE, Aura T, Zahemszky A, Nikander P, Ott J (2011) Forwarding anomalies in bloom filter-based multicast. In: Proc. IEEE INFOCOM, 2011, pp 2399–2407
15. The network simulator – ns-2. http://www.isi.edu/nsnam/ns
16. Tian X, Cheng Y (2012) Loop mitigation in bloom filter based multicast: a destination-oriented approach. In Proc. IEEE INFOCOM
17. Tian X, Cheng Y, Liu B (2009) Design of a scalable multicast scheme with an application-network cross-layer approach. IEEE Trans Multimedia 11:1160–1169
18. Tian X, Cheng Y, Liu B (2010) A fast-join mechanism for inter-domain multicasting. In: Proc. IEEE GLOBECOM, 2010
19. Wikipedia. http://en.wikipedia.org/wiki/Bin_packing_problem

Chapter 5
AOM-Assisted Zapping Acceleration for IPTV[1]

Abstract Channel zapping time is a critical quality of experience (QoE) metric for IPTV service. Recently, zapping acceleration scheme based on time-shifted sub-channels (TSS) is introduced [1]. In this chapter, we develop a systematical analysis of TSS-based service model to understand its fundamental properties from a theoretical perspective; moreover, we propose an AOM-Assisted Zapping Acceleration (AAZA) scheme to realize TSS-based model, which improves the performance of IPTV systems. Section 5.1 briefly describes the channel zapping issue of IPTV systems and existing solutions. In Sect. 5.2, the basic principles of the TSS-based service model is presented. Section 5.3 analyzes the TSS-based model from a theoretical perspective. We show that there exists an optimal subchannel data rate which minimizes the redundant traffic transmitted over subchannels; moreover, we reveal the *start-up effect*, where the original operation pattern in [1] could violate the performance bound of the TSS-based model. With a convenient solution to the start-up effect proposed, we rigorously prove that the bounded zapping time is guaranteed in the TSS-based model. In Sect. 5.3, we describe the AAZA scheme to realize TSS-based model, which achieves the seamless subscriber migration from the subchannel to the main channel, without any control message exchange over the network. The subchannel selection in AAZA is independent of the zapping request signaling time, resulting in improved robustness and reduced messaging overhead. In Sect. 5.5, we implement AAZA in ns-2 and multicast an MPEG-4 video stream over a practical network topology. Extensive simulation results are presented to demonstrate the validity of our analysis and AAZA scheme. Section 5.6 summarizes this chapter.

[1] ©[2012] IEEE. Portions reprinted, with permission, from [17].

X. Tian and Y. Cheng, *Scalable Multicasting over Next-Generation Internet: Design,*   103
Analysis and Applications, DOI 10.1007/978-1-4614-0152-0_5,
© Springer Science+Business Media New York 2013

5.1   Zapping Acceleration in IPTV Systems

IPTV systems deliver the TV program as a compressed multicast data stream (e.g., MPEG-2/4 and H.264 [18]) over IP-based networks, where the stream is a series of groups of pictures (GOPs) and play-out can only start with an I-frame at the beginning of each GOP. The IPTV channel zapping is the act of leaving a stream and joining in another. The duration it takes for the picture of the new TV channel to start displaying since the zapping request has been issued is *zapping time*, which is a critical quality of experience (QoE) metric for IPTV systems. As illustrated in Fig. 5.1, the zapping request possibly occurs at any time in a GOP of the new channel stream the time between the arrival of the request and the first I-frame (*first I-frame delay, FID*) could be up to a few seconds, which is a significant contributor to the IPTV channel zapping time [1].

A number of techniques have been proposed recently to mitigate the IPTV zapping time, most of which are realized with the auxiliary stream that starts with an I-frame. The set-top-box (STB) trying to join in a new channel first subscribes to the auxiliary stream and then migrates to the main multicast stream when enough data are accumulated in the STB play-out buffer. There have been four basic approaches to exploit the auxiliary stream (Table 5.1):

- *Approach 1:* Unicast a full-quality boost stream replicated from the previously stored stream data for each zapping request [5, 18]. The approach can impose significant resource demands on networks and the streaming server.
- *Approach 2:* Generate the low-quality stream, which is composed of just I-frames [11] or several low-resolution channels [6, 12], to accompany the regular channel stream. Such mechanisms may incur noticeable picture inconsistency at each channel zapping act.

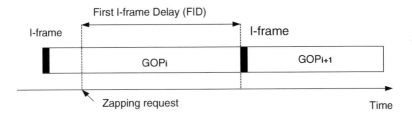

Fig. 5.1 First I-frame delay in IPTV channel zapping

Table 5.1 Features comparison of zapping acceleration solutions

Features	Approach 1	Approach 2	Approach 3	Approach 4	TSS-based model
Server burden	High	Medium	Medium	Medium	Low
Network burden	High	Medium	High	Medium	Low
Picture inconsistency	No	Yes	No	No	No
New codec	No	No	No	Yes	No

- *Approach 3:* Join additional multicast groups along with the current channel group, where the additional one could be simply the adjacent multicast group to the current channel[10], or the most likely next channel predicted by users' channel selection behaviors [7,8]. These solutions require extra streams delivered to the home gateway for each channel being watched, thus incur expensive bandwidth cost in the access network.
- *Approach 4:* Encode a low-quality auxiliary stream with frequent I-frames into the regular stream for each channel [2, 3]; however, the STB has to be equipped with extra codec and the video has to be recoded.

Recently, MAZA [1], a channel zapping acceleration scheme based on the time-shifted subchannels (TSS), is proposed. In MAZA, a TV channel is companied with several subchannels that are spaced by T time units, and each subchannel is a full-quality replica of the main channel media stream. The TSS-based scheme can guarantee a bounded FID and provide the following outstanding advantages over existing solutions. First, the subchannel stream is delivered through multicast to subscribers upon request, thus consumes less bandwidth resources in contrast to the unicasting boost stream or additional group pre-joining method. Second, there is no picture inconsistency incurred by the low-quality accompanying stream during the transition from the subchannel to the main channel. Third, there is no new codec introduced to accommodate the auxiliary stream encoded into the regular channel.

Despite the remarkable advantages, the fundamental properties and performance of the TSS-based service model are not yet well understood from a theoretical perspective. Moreover, the current implementation of TSS-based model utilizes IP multicast, with each channel and its associated subchannels implemented as separate multicast groups. The per-group-based multicasting is well known as suffering from the scalability issue in a large distributed system and in particular can incur frequent control messages for the TSS-based channel zapping.

In following sections, we develop a systematical framework to mathematically analyze and understand the fundamental properties of TSS-based service model. Specifically, we show that there exists an optimal subchannel data rate which minimizes the redundant traffic transmitted over subchannels; moreover, we reveal the *start-up effect*, by which the basic channel turn-on policy in [1] could violate the FID bound. We thus introduce an augmented channel turn-on policy for guaranteed FID performance. Moreover, we give a rigorous proof that the TSS-based model could ensure a FID bound of T under the augmented turn-on policy.

Furthermore, we propose an application-oriented-multicast (AOM)-Assisted zapping acceleration (AAZA) scheme to implement the TSS-based service model. The AOM [15, 16] scheme proposed in the previous chapters makes each packet carry explicit destinations of receivers, instead of an implicit group address as in IP multicast, to facilitate multicast forwarding. AOM is an excellent match to the TSS-based model, with which AAZA achieves the seamless subscriber migration from the subchannel to the main channel, incurring no control message exchange over the network. The subchannel selection is simply accomplished at the data source node, independent of the zapping request signaling time, thus resulting in improved robustness and reduced messaging overhead of AAZA. We implement AAZA in ns-2 and multicast an MPEG-4 video stream over a practical network topology. Extensive simulation results are presented to demonstrate the validity of our analysis and AAZA scheme.

We first give a brief introduction to MAZA in the next section, so that the following analysis can be understood.

5.2 MAZA

TSS-Based Service Model of MAZA. In order to constrain the channel zapping time by reducing the FID, the TSS-based model was recently proposed in MAZA scheme [1]. Specifically, each TV channel is accompanied by several time-shifted subchannels (sCHs) that are generated through replicating the main channel (mCH) media stream as illustrated in Fig. 5.2. There are $X = \lceil \frac{s}{T} \rceil^2$ sCHs coexisting with the mCH, where T is the time space between two adjacent sCHs and s is the size (in time units) of the largest GOP in the video stream. Every T time units, an I-frame will be sent over a sCH. At any time, a joining STB can always find an appropriate sCH, which will send an I-frame within T time units. With TSS-based model, the observed FID can be bounded by T.

However, if sCHs are always active, they will consume network resources continuously. To save network resources, the TSS-based model provides the following two desirable functionalities. First, a user joins a sCH will migrate to the mCH as soon as possible, and any sCH on which no users are listening can be deactivated. Second, a sCH transmits data only when there is a user who will need its service.

For the first functionality, the sCH intuitively only needs to be configured a higher data rate considering mCH's head start. The sCH subscribers can change to join in the mCH after the sCH catches up with the mCH, and the sCH can be deactivated then. The problem is, if the X sCHs are deactivated one after another, there will

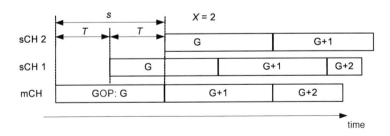

Fig. 5.2 TSS-based service model

---

[2]When $\frac{s}{T}$ is an integer, the $\frac{s}{T}$th sCH lags behind the mCH by exactly a GOP and aligns to the mCH at an I-frame. However, this sCH is necessary in order to guarantee the delay bound of T, to be proved in Sect. 5.3.

be only mCH in the system, and the FID of new joining subscribers cannot be bounded. The TSS-based model adopts dynamic sCH operations to deal with this issue. Imagine that there are infinite number of sCHs that are spaced by T and configured a higher data rate than the mCH. As we only need X sCHs to guarantee the FID bound, the rest of the sCHs can be imagined as invisible. Suppose the sCH 1 in Fig. 5.2 catches up with the mCH, we could make sCH 3 visible in order that the new comers could be taken care of by sCH 2 and 3. In this way, there are always X sCHs to ensure the T-shift property, and sCH subscribers will eventually migrate to the mCH.

For the second functionality, TSS-based model does not need sCHs to transmit data all the time. In fact, only some meta information about sCHs needs to be maintained. For example, which sCHs should be visible? Which subscribers are listening on a given sCH? Which part of the media stream should be delivered over a sCH that has users listening on? With these information, TSS-based model can always duplicate correct part of the streaming data from the mCH and dispatch them to corresponding sCH users only when requested.

IP-Multicast-Based Implementation of MAZA. MAZA implements the TSS-based service model in the streaming server called zapping accelerator (ZA), which delivers the streaming data to STBs with IP multicast. In MAZA, each mCH or sCH is assigned an IP multicast address, and the STB is pre-configured to associate the IP multicast address with the mCH or sCH index. There is a *meta-channel* in ZA broadcasting to all STBs in advance, which specifies the sCH offering the earliest I-frame for every GOP. After the STB receives the meta-channel information, it can decide which sCH to join in, with considering the time it takes for the zapping request to reach ZA (signaling delay). When the chosen sCH catches up with mCH, the ZA notifies the STB to migrate to mCH and continues keeping the sCH on to have enough time for the STB to join in mCH. The IP-multicast-based implementation can be affected by the signaling delay and incur frequent control messages. These issues will be eliminated with the facilitation of DOM.

5.3   Theoretical Analysis of the TSS-Based Model

This section presents a fundamental theoretical study of the TSS-based service model. We first summarize in a more understandable manner the basic operational properties of the TSS-based model according to the results in [1]. By examining the lifetime of a sCH, we reveal that an optimal sCH data rate exists, which leads to the least amount of redundant traffic. We then investigate the start-up effect, by which the basic channel turn-on policy in [1] could violate the FID bound. We thus propose an augmented channel turn-on policy for guaranteed FID performance. Moreover, we give a rigorous proof that the TSS-based model could ensure a FID bound of T under the augmented turn-on policy.

5.3.1 Operation Pattern of TSS-Based Model

The essence of the TSS-based system is to maintain the proper time-shifting relationship among sCHs, considering the dynamics due to channel merging and creation. In accordance with [1], the core operation pattern is the sCH turn-on policies. *sCH Turn-On Policies*[3]

1. *For the first X sCHs, they are created and turned on one after another spaced by T.*[4]
2. *For $i > X$, sCH i is turned on when sCH $i - X$ catches up with mCH.*

Specifically, the creation and turn-on times of the first X sCHs are the same, which conserve the T-shift property,

$$C_i = C_m + iT \quad 1 \leq i \leq X, \tag{5.1}$$

where C_i is the creation time of sCH i and C_m is the creation time of the mCH.

A higher sCH rate R compared to mCH rate is purposely configured so that an active sCH can later catch and merge into the main channel to reduce the redundant traffic. It is not difficult to calculate that the sCH i will catch mCH at $C_i + iT/\Delta$, where $\Delta = \frac{R}{r} - 1$, considering that sCH i lags behind the mCH with data volume iTr when it is created. A channel merging event will logically turn on a new sCH, so that the TSS system always logically maintains X sCHs. The term "logically maintain" means that the time-shifting relationship of each sCH is calculated, but the actual data transmission (i.e., activation) of a sCH depends on the service request. Such a sCH service model could achieve the balance between guaranteed FID performance and controlled traffic redundancy.

For the the convenience of analysis, we sequentially index all the sCHs that will be involved. Since the neighboring sCHs have a clear T-shifting relationship, we can consider that those sCHs turned on later (i.e., sCHs $i > X$) are virtually created at $C_i = C_m + iT, i > X$. Note that when a sCH i catches mCH, it will turn on the sCH $i + X$. Let V_{i+X} denote the turn-on time of sCH $i + X$. We have

$$C_{i+X} = C_m + (i+X)T, \quad i \geq 1 \tag{5.2}$$

$$V_{i+X} = C_i + \frac{iT}{\Delta}, \quad i \geq 1. \tag{5.3}$$

For the convenience of implementation, it is important to determine the packet index for transmission when a sCH is turned on under the time-shifting constraint. Consider that the packet for transmission over sCH $i + X$ at the turn-on moment

---

[3]See Sect. IV-C of [1].

[4]Upon creation/turning on, the sCHs do not necessarily transmit data. Only the timing relationship is maintained. A sCH will physically transmit data only when it has subscribers.

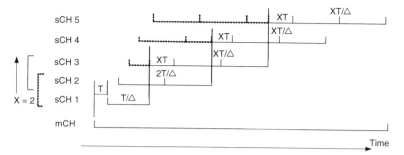

Fig. 5.3 Operation pattern of TSS-based model

V_{i+X} was transmitted on mCH θ_{i+X} time units ago. Since the sCH $i+X$ lags behind the sCH i with time XT and sCH i just catches mCH at V_{i+X}, thus,

$$\theta_{i+X} = \frac{XTR}{r}. \tag{5.4}$$

The key observation underpinning (5.4) is that the data rates of sCHs and mCH are different.[5] At V_{i+X}, although sCH $i+X$ lags behind the sCH i with time XT and sCH i is transmitting the same packet as mCH, we need to trace back more than XT time units over mCH to retrieve the packet that should be transmitted over sCH $i+X$ at the moment.

We consider the *lifetime* of sCH $i+X$ since it is turned on till it catches mCH. According to the virtual creation time (5.2), it can be seen that sCH $i+X$ will catch mCH at $C_{i+X} + (i+X)T/\Delta$. Let D_{i+X} denote the life time of sCH $i+X$, and we have

$$D_{i+X} = C_{i+X} + (i+X)T/\Delta - V_{i+X}$$

$$= XT\left(1 + \frac{1}{\Delta}\right) \quad i \geq 1. \tag{5.5}$$

Note that the TSS system has a regular pattern in merging and turning-on sCHs. The above analysis generally applies to every sCH dynamically turned on (Fig. 5.3).

5.3.2 Optimal Subchannel Data Rate

In the TSS model [1], all the packets transmitted over a sCH are replicated data of a corresponding portion of mCH, which are the traffic redundancy. Let $B(R)$ denote

---

[5]The parameter θ_{i+X} in [1] is actually the initial time gap between sCH $i+X$ and sCH i, instead of mCH.

the amount of traffic delivered over a sCH during its lifetime. Based on (5.5), we have

$$B(R) = RXT \left(1 + \frac{1}{\Delta} \right)$$

$$= RXT + \frac{RXT}{\Delta}. \tag{5.6}$$

To minimize the redundant traffic, we can determine the optimal sCH data rate R^* in the range (r, ∞) according to

$$\frac{d}{dR} B(R) \mid_{R=R^*} = 0, \tag{5.7}$$

which gives

$$R^* = 2r. \tag{5.8}$$

It is interesting to interpret why there exists an optimal sCH rate and why the optimal rate is $2r$. Based on the analysis above, we can see that the total amount of traffic delivered during the lifetime of a sCH consists of two parts: the *initial lagging traffic* at the moment when the sCH is turned on and the *extra catching traffic* during the catching procedure. The sCH data rate R has contradicting effects on the two parts. The analysis in Sect. 5.3.1 shows that the initial lagging traffic is XTR when a sCH is turned on by a sCH merging event in a general context, which is monotonically increasing with R. Equation (5.6) shows that the extra catching traffic is $RXT/\Delta = \frac{XTr}{1-\frac{r}{R}}$ (with $\Delta = \frac{R}{r} - 1$), which is monotonically decreasing with R. The optimal value $R^* = 2r$ balances the two contradicting effects, where $R^*XT = R^*XT/\Delta$ and the sum of these two items is minimized.

5.3.3 Start-Up Effect

The *start-up effect* occurs when the sCHs start to be turned on by the channel merging events: Simply turning on the sCHs according to (5.3) and starting packet transmission according to the time relationship (5.4) may lead to the FID greater than T. Consider a scenario shown in Fig. 5.4a, where the largest GOP size $s = 4T$ ($X = 4$), $R = 2r$, and the time shift is T. Each section of line represents the duration of a sCH, with its index labeled. The sCHs 1–4 are turned on every T time units. When sCH 1 catches mCH, the sCH 5 should start transmitting, and the first packet should be on mCH $8T$ time units ago according to (5.4). However, since mCH just started $2T$ ago, the sCH 5 at best can start from the first packet on mCH. The same can happen to sCHs 6–8. Suppose that an I-frame, represented as a black rectangle, is transmitted at $4T$ over mCH. The time points where it reappears over existing sCHs can be easily identified. For example, sCH 3 starts from packet 0 and the data

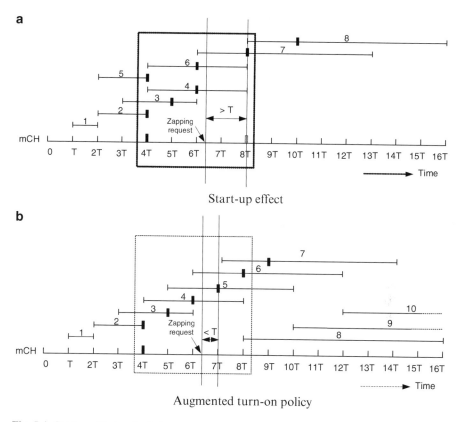

Fig. 5.4 Start-up effect and solution

rate is twice as that of mCH; thus, the I-frame will reappear on sCH 3 $2T$ after its creation. Here we assume the first two GOPs are s long. The grey rectangle in the figure is the I-frame of the second GOP. If a zapping request arrives at some time between $6T$ and $7T$, the earliest I-frame will be seen is at $8T$ as illustrated in Fig. 5.4a; thus, the FID is greater than T.

The start-up effect particularly impacts the sCHs $i + X$, $i \geq 1$. To avoid the situation that a sCH tends to send a packet even before it is generated, a constraint for turning on sCHs by channel merging event should be the channel turn-on time should not be earlier than the channel creation time, that is,

$$V_{i+X} \geq C_{i+X}. \tag{5.9}$$

After some manipulation, we have

$$i \geq X\Delta.$$

Therefore, the channel merging event can turn on sCHs with index

$$i + X \geq X(\Delta + 1). \tag{5.10}$$

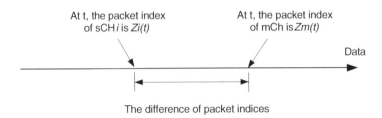

Fig. 5.5 Packet indices difference between sCh i and mCH

The constraint (5.10) indicates that the turn-on policies could be augmented.
Augmented sCH Turn-On Policies

1. *For the first $\lceil X(\Delta + 1)\rceil - 1$ sCHs, they are created and turned on one after another spaced by T.*
2. *For $i \geq \lceil X \cdot (\Delta + 1)\rceil$, sCH i is turned on when sCH $i - X$ catches up with mCH.*

Applying the augmented policies to the system illustrated in Fig. 5.4a, the start-up effect will not happen, and the model guarantees a bounded FID, which is shown in Fig. 5.4b. It is worth mentioning that the augmented sCH turn-on policies will not impact the results presented in (5.4)–(5.8) as long as the sCH turn-on policy according to the channel merging event starts to take effect.

5.3.4   Delay Bound Guarantee

Although it is intuitively true that using X $(= \lceil \frac{s}{T}\rceil)$ T-shifted sCHs to replicate mCH can ensure a FID bound of T time unites, it is meaningful to give a rigorous theoretical proof of such FID bound for the general scenario where the GOPs may have heterogeneous sizes and the augmented sCH turn-on policy is applied.

For the convenience of presentation, we term the regular T-shifted turn-on policy as policy 1 and the policy upon the channel merging event as the policy 2. Notice that for the sCHs turned on under policy 1, their turn-on times are the same as their creation times and thus T-spaced. However, the turn-on times of neighboring sCHs under the policy 2 are not T-spaced. Based on (5.2) and (5.3), it can be checked that for a sCH $i(\geq \lceil X(\Delta + 1)\rceil)$ being turned on under policy 2, the gap between successive turn-on moments is $V_{i+1} - V_i = T + \frac{T}{\Delta}$. Regarding the data transmission, it is important to check whether the T-shifting property can still maintain under the policy 2. We have the following lemma regarding the time-shifting property (Fig. 5.5).

Lemma 5.1. *For any two neighboring sCHs i and $i + 1$, either under the turning-on policy 1 or policy 2, a packet delivered over sCH i will be delivered over sCH $i + 1$ at time $t + T$.*

Proof. We prove the lemma by establishing the time lag relationship between a given sCH and mCH. Let $H_i(t)$ denote the time lag of sCH i regarding mCH, that is, the packet being transmitted over mCH at t will be retransmitted over sCH i at time $t + H_i(t)$.

We first examine the time-shifting relationship for two neighboring channels turned on under policy 2. At any time t, the index of the packet being transmitted over mCH is $Z_m(t) = r \cdot (t - C_m)$. The sCH turn-on operation is that when the sCH $i - X$ merges into the main channel, the sCH i is turned on (at time V_i) and starts transmission at rate R from the packet with an index XTR earlier than that over mCH at the moment (referring to the discussion in Sect. 5.3.3). Under such an operation, at time t, the packet index over the sCH i is $Z_i(t) = r(V_i - C_m) - XTR + R(t - V_i)$. We can then find the time lag

$$H_i(t) = \frac{Z_m(t) - Z_i(t)}{R} = \left(1 - \frac{r}{R}\right)(V_i - t) + XT,$$

$$i \geq \lceil X(\Delta + 1) \rceil. \tag{5.11}$$

Based on (5.11), the time gap between sCH i and sCH $i + 1$ can be obtained as $H_{i+1}(t) - H_i(t) = \left(1 - \frac{r}{R}\right)(V_{i+1} - V_i) = T$ (recall that $V_{i+1} - V_i = T + \frac{T}{\Delta}$ and $\Delta = \frac{R}{r} - 1$).

For a sCH i ($\leq \lceil X(\Delta + 1) \rceil - 1$), we can easily see that the packet index at t is $Z_i(t) = R(t - C_i)$. The packet index over mCH is still $Z_m(t) = r(t - C_m)$. We can then compute

$$H_i(t) = \frac{Z_m(t) - Z_i(t)}{R} = \left(1 - \frac{r}{R}\right)(C_m - t) + iT,$$

$$i \leq \lceil X(\Delta + 1) - 1 \rceil. \tag{5.12}$$

Based on (5.12) and $C_i = C_m + iT$ for $i \leq \lceil X(\Delta + 1) - 1 \rceil$, we can have the time gap between neighboring sCHs is $H_{i+1}(t) - H_i(t) = T$.

Finally, we consider the gap between channel $\lceil X(\Delta + 1) \rceil - 1$ and channel $\lceil X(\Delta + 1) \rceil$, turned on by policy 1 and policy 2, respectively. For convenience, denote $K = \lceil X(\Delta + 1) \rceil$. By applying $V_K = C_{K-X} + \frac{(K-X)T}{\Delta}$ (the merging moment of sCH $K - X$) to (5.11) and $C_{K-1} = C_m + (K - 1)T$ to (5.12), we can also have $H_K(t) - H_{K-1}(t) = T$ after some manipulation.   □

With Lemma 5.1, we can then prove the FID bound of T time units. Specifically, we have the following theorem:

Theorem 5.1. *In the TSS-based service model, there is at least one channel (including mCH and sCHs) which guarantees that an I-frame will appear within T time units after the channel zapping request arrives.*

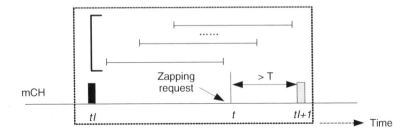

Fig. 5.6 Delay bound guarantee of TSS-based model

Proof. Consider that the channel zapping request arrives at t; the latest I-frame before the zapping request was delivered over mCH at $t_l \leq t$, and the newest I-frame after the zapping request is at $t_{l+1} > t$. Since the maximum GoP is s, we have $t - s < t_l \leq t$ and $t_{l+1} - t_l \leq s$.

Let A denote the current time lag of the earliest sCH (the active sCH with the smallest channel index) regarding mCH; we have $0 < A \leq T$. Under the turn-on policy 1, it is clear that sCH 1 starts with a time lag of T, but the time lag will reduce because the sCH rate is higher than mCH rate (i.e., $R > r$). Under the turn-on policy 2, upon the channel merging event, the existing earliest sCH has a time lag of T at that moment and again the time lag will reduce along with the time because $R > r$. Given a time t, we can index the current earliest sCH as sCH 1 without loss of the generality, and thus $H_1(t_l) = A$. Since all the neighboring sCHs have a time lag relationship of T according to Lemma 5.1, we then have

$$H_i(t_l) = A + (i-1)T, \quad i \in [1, X] \tag{5.13}$$

for all the X sCH that are currently logically active. For convenience, we denote mCH as $i = 0$. To prove the theorem, we examine two cases as shown in Fig. 5.6.

Case 1: Zapping request starts being served from frame I. If $t - t_l \leq s - T$, we can show that a channel index $i \in [0, X]$ exists so that

$$t \leq t_l + H_i(t_l) \leq t + T. \tag{5.14}$$

From (5.13) and (5.14), we have

$$t \leq t_l + A + (i-1)T \leq t + T, \tag{5.15}$$

and thus

$$\frac{t - t_l - A}{T} + 1 \leq i \leq \left(\frac{t - t_l - A}{T} + 1\right) + 1. \tag{5.16}$$

Note that $0 \leq t - t_l \leq s - T$ and $-T \leq -A < 0$, we can see that

$$0 \leq \frac{t - t_l - A}{T} + 1 < \frac{s}{T}. \tag{5.17}$$

Recall that $X = \lceil \frac{s}{T} \rceil$, so the results of (5.16) and (5.17) guarantee that a channel index $i \in [0, X]$ exists.

Case 2: Zapping request starts being served from frame $I + 1$. If $t - t_I > s - T$, the zapping request will be served with mCH, starting from the frame $I + 1$. Note that $t_{I+1} - t_I \leq s$; thus, we have $t_{I+1} - t < T$.   □

5.4 AAZA: AOM-Assisted Zapping Acceleration

In this section, we first describe the design of AAZA, which implements the TSS-based model and integrates it into the IPTV infrastructure. Then the properties of AAZA are analyzed.

5.4.1 TSS-Based Model Facilitated by AOM

The AAZA server can be deployed at the provider network to accelerate the channel zapping. TV channels requiring zapping acceleration are first streamed to the AAZA server, where the time-shifted sCHs for each channel are generated. To implement the TSS-based service model, the AAZA server maintains two fundamental data structures, *subchannel states table* (SST) and *AOM destinations cache* (ADC), as in Fig. 5.7. We here consider the single group case (GID$_1$) for the convenience of presentation.

Subchannel Selection. SST maintains information of $X + 1$ channels that are logically turned on. For each sCH, the index number, turn-on time, and the time gaps between the sCH and mCH at t_I are recorded, where t_I is the transmission time of the latest I-frame over mCH. Initially, the SST table contains X sCHs spaced by T. The SST will be updated according to the augmented sCH turn-on policies as analyzed in Sect. 5.3.3. The sCH with the lowest index will first outspace mCH, as the shadowed row of SST in the left side of Fig. 5.7a. Then, the table entry will be replaced by a new sCH as shown in the left side of Fig. 5.7b. For the new sCH added to the SST, its turn-on time is determined by the sCH turn-on policies, while $H_i(t_I)$ can be computed using (5.11) or (5.12), based on the value of the sCH index. After the sCH index goes beyond $\lceil X \cdot (\Delta + 1) \rceil$, the SST will be updated every $T + \frac{T}{\Delta}$ time units as analyzed in Sect. 5.3.3.

When a zapping request is received, the AAZA server first checks the SST to find the sCH i^* that will provide the earliest I-frame, as illustrated in Algorithm 1.

Note that Algorithm 1 corresponds to the two cases we proved in Theorem 5.1; thus, the AAZA server can always find at least one of the channels, which will transmit an I-frame within T. The physical meaning of Algorithm 1 is illustrated in Fig. 5.8 for better understanding. In the example, when a zapping request is received at t, the latest I-frame was transmitted over mCH at t_I, and this I-frame will be sent over each of the 3 sCHs at $H_i(t)$. Figure 5.8a shows an example of the first case in Algorithm 1, where the zapping request arrives so late that it misses the I-frame

a

Sub-Ch States Table for GID1

CH #	Vi	Hi(l)
0 (mCH)	Cm	n/a
1	Cm+T	T
2	Cm+2T	2T
...	...	...
X	Cm+XT	XT

AOM Dest. Cache for GID1

CH #	Subscribers
0 (mCH)	R1, R2
1	R3, R4
2	R5
...	...
X	R6, R7, R8

Initial states

b

Sub-Ch States Table for GID1

CH #	Vi	Hi(l)
0 (mCH)	Cm	n/a
X+1	C1+T△	h
2	Cm+2T	2T
...	...	...
X	Cm+XT	XT

AOM Dest. Cache for GID1

CH #	Subscribers
0 (mCH)	R1, R2, R3, R4
X+1	R9
2	R5
...	...
X	R6, R7, R8

Updated states

Fig. 5.7 Data structure of AAZA

Algorithm 1: sCH selection

Input: t_I - the time the latest I-frame before the zapping
request was delivered over mCH;
t - the time a zapping request is received;
Output: i^* - the sCH to provide the earliest I-frame;
if $\forall i \in [1, X], (t_I + H_i(t_I) - t) < 0$ **then**
  | $i^* = 0$;
else
  | $i^* = argmin_i\{(t_I + H_i(t_I) - t) \geq 0\}$;
end

---

transmitted over all 3 sCHs; therefore, the earliest I-frame must be available over mCH. Figure 5.8b shows an example of the second case of Algorithm 1, where sCH 1's corresponding I-frame has been missed but sCH 2 has the I-frame on the way and the waiting time must be bounded by T.

Data Transmission Over Subchannels. Among those "logically on" sCHs, only the sCHs with subscribers are actually transmitting data packets. The transmitted packets over the sCHs are replicated from the appropriate portion of mCH. At any time t, the packet being transmitted over sCH i should have been transmitted over mCH at corresponding time $P_i(t)$. Algorithm 2 shows how to identify the $P_i(t)$.

We here explain the physical meaning of Algorithm 2. If $t \leq V_i$, meaning the sCH i should wait to be turned on, so sCH i should retrieve the data transmitted over mCH θ_i time units before sCH i is turned on. The initial time gap between sCH i and mCH, θ_i differs according to the index of the sCH as indicated by the augmented turn-on policies in Sect. 5.3.3.

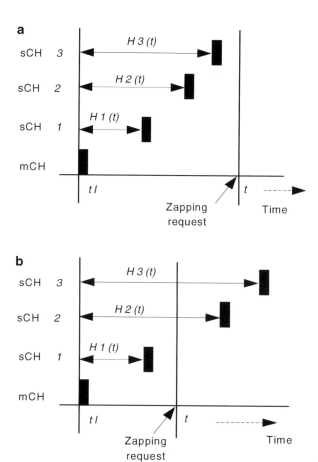

Fig. 5.8 Subchannel selection

If $t > V_i$, the sCH i has been turned on, its starting packet index is $r(V_i - \theta_i)$, and the packet index becomes $r(V_i - \theta_i) + R(t - V_i)$ at t; therefore, the packet with this index has been transmitted over mCH at

$$P_i(t) = \frac{r(V_i - \theta_i) + R(t - V_i)}{r} = \Delta(t - V_i) + t - \theta_i.$$

Seamless Migration to Main Channel. ADC is used to record subscriber addresses for the mCH and active sCHs and for sCH subscribers to seamlessly migrate to mCH. ADC is initialized by grouping subscribers that select the same sCH as shown in the right side of Fig. 5.7a, where mCH is also included. For a given sCH, if it catches mCH, that is, the play-outs over the sCH and mCH are the same, the subscribers of the sCH will be merged into the set of main channel subscribers. Refer to the right side of Fig. 5.7b, sCH 1 catches mCH, so the original subscribers $R3, R4$ of sCH 1 merge into mCH, as highlighted in the shadowed area. The channel index

---

Algorithm 2: Data to be transmitted over sCH i

Input: t - the time;

$\qquad V_i$ - turn-on time of sCH i when sCH $i - X$

$\qquad$ catches mCH;

$\qquad \theta_i$ - initial time gap between sCH i and mCH;

Output: $P_i(t)$ - at any time t, the packet being transmitted over sCH i should have been

$\qquad$ transmitted over mCH at time point $P_i(t)$;

if $i < \lceil X(\Delta + 1) \rceil$ **then**

$\qquad | \quad \theta_i = iT$;

else

$\qquad | \quad \theta_i = \frac{XTR}{r}$;

end

if $t \leq V_i$ **then**

$\qquad | \quad P_i(t) = V_i - \theta_i$;

else

$\qquad | \quad P_i(t) = \Delta(t - V_i) + t - \theta_i$;

end

---

of ADC is for AAZA server to retrieve correct packet payloads, as different sCHs are transmitting different parts of the video as indicated by the output of Algorithm 2. The addresses of the subscribers (Ri) will be encapsulated into the multicasting packet labeled with group ID GID$_1$. AOM data forwarding protocol will deliver the packet to corresponding destinations. The subscriber migration needs no message exchange over the network.

5.4.2   Performance Analysis

AAZA and MAZA both implement the TSS-based service model, while the fundamental difference is that AAZA selects sCHs at the streaming server while MAZA does it at the STBs.

Signaling Delay Effect. The time taken for an STB to join in a multicasting group (signaling delay) can possibly affect the FID of MAZA but will not affect AAZA. In MAZA, the meta-channel information is broadcast to STBs in advance. After the STB receives the meta-channel information, it has to take the signaling delay into account to select the sCH to join in. Consider the scenario illustrated in Fig. 5.9; the appearance of I-frame I is notified to the STB J time units in advance, where J is the signaling delay predicted. If a zapping request arrives at the STB at t, the STB should find out which sCH will provide the earliest I-frame at $t + J$ time units. In Fig. 5.9, the STB will apparently take sCH 1. However, as it is difficult to predict the traffic condition of access networks, the joining message can be very likely delayed for a longer time than J and miss the time when the selected sCH is transmitting the wanted I-frame. This happens as the late joining message case shown in Fig. 5.9, and the arrived joining message has to wait up to XT time units for the next I-frame.

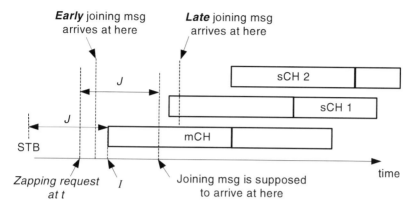

Fig. 5.9 Signaling delay effect

The signaling delay could be very short; however, the major issue is that MAZA requires the accurate prediction of the signaling delay for sCH selection, which is quite difficult in practice. For example, even if the signaling delay can be bounded, it is also possible that the joining message arrives at the MAZA server earlier than expected. In the early joining message case of the Fig. 5.9, the early arrived joining message has to wait more than T to get the first I-frame. If the STB had known the joining message could take such short time to reach the MAZA server, it would have joined in mCH, instead of sCH 1. In contrast, AAZA selects sCHs at the streaming server after the joining message arrives, which is totally independent of the signaling delay.

Control Messages Overhead. AAZA incurs less control messages overhead than MAZA does. The control message exchange process for a zapping act under MAZA is illustrated in Fig. 5.10. MAZA server broadcasts the meta-channel information for each GOP of every TV channel provisioned. If the STB chooses a sCH upon a zapping request, it is in the sCH until the sCH merges into mCH. At the moment of merging, the MAZA server sends a notification to indicate the STB to join in mCH. To this end, the STB sends a leaving message to unsubscribe the sCH and then sends another joining message to migrate to mCH [12, 32]. Frequent messages exchange incurs higher probability of system errors. In contrast, the AAZA only needs one joining message to switch to the new channel, and the migration from sCH to mCH is performed without any message exchange over the network. Moreover, the MAZA meta-channel broadcasts to STBs each time an I-frame is sent over mCH, while AAZA needs no meta-channel broadcasting. In AAZA, DOM constructs the source-based multicast tree using destination-specific forwarding states [15,16]. The zapping operation from an STB does not need to tear down the multicasting tree branch as long as the new channel is still provided by the same data source.

Flexibility. The channel management in AAZA is more flexible than that in MAZA. MAZA assigns an IP address for each TV channel and corresponding sCHs. Since the IP addresses for multicasting are not always continuous, each STB has to be pre-configured an address pool for associating each channel index with its IP address.

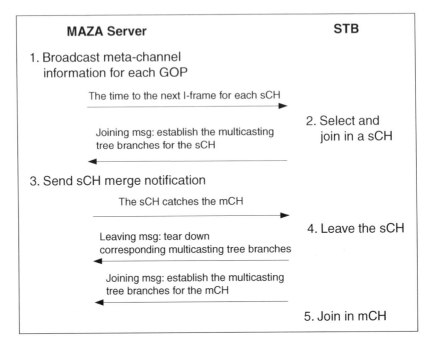

Fig. 5.10 Control message exchange in MAZA

If the service provider wants to reallocate the multicast addresses for sCHs, all STBs have to be reconfigured. For example, if the service provider wants to reconfigure T, as T will impact the number of sCHs (X) maintained by the MAZA server, some sCH addresses could be recollected and distributed to other TV channels. The STB reconfiguration in MAZA will be complicated in this case. AAZA can provide a flexible channel management scheme. The fundamental reason is that AOM decouples the forwarding from group identifiers and provides to allocate group identifiers locally at the data server [15, 16]; therefore, STBs can receive correct data packets without even knowing identifiers of sCHs.

Scalability. A major concern for AAZA is that the multicasting packet may not be able to encode all the subscribers' IP addresses if there are a large number of subscribers, which may hinder AAZA from scaling up. In fact, if an STB requests to switch to an existing channel that is already being received by other STBs in the receiver domain, the request can be served by the local AAZA server for the domain. Otherwise, the request has to be sent to the AAZA server located in the core network. For both cases, the scalability issue can be handled.

Recall the typical wireline broadband access network architecture as illustrated in Fig. 2.14; the STB initiates the channel zapping request as an IGMP request for joining/leaving corresponding groups [12, 32]; the request passes through the routing gateway (RG), digital subscriber line access multiplexer (DSLAM), and will finally reach the broadband services router (BSR) acting as the gateway into the backbone network. The technique of *local content insertion* [4] can be applied to provision a fast local channel zapping process, where a local AAZA server can be connected to the BSR. The AAZA server maintains all channels that are being watched by the local domain. If the requested channel is being viewed by existing subscribers, the zapping request will be served by the local AAZA server. For a given channel, the corresponding sCH subscriber will eventually merge into the mCH, and 90% of receivers tend to be interested in the top 10% of the channels according to the study on IPTV users behavior [13, 19]; therefore, the subscribers' IP addresses list of each popular channel will be considerably long. In order to avoid the expense of this lengthy enumeration in the DOM multicasting header, we can use the *multicast push* model [43], where a channel will be broadcast if the number of its subscribers is beyond certain proportion of the total subscribers in the domain.

If the STBs request new channels that are not maintained in the local AAZA server, the BSR needs to unicast the aggregated membership information toward corresponding AAZA server in the core network to establish the multicasting tree. Consequently, the addresses encoded into the AAZA multicasting packet are the BSRs' addresses. The scalability of DOM in this inter-domain scenario is described in detail in [15, 16]. Particularly, a large number of BSRs still can incur bandwidth overhead in AAZA, as multiple data packets need to be generated to cover all BSRs, with each covers a subset of the BSRs. However, the bandwidth overhead of AAZA is reasonably low and close to that of MAZA. Section 5.5.6 gives the simulation results for comparing the bandwidth overhead under AAZA with that under MAZA.

5.5 Simulation Results

We use ns-2 [14] simulation results to verify our theoretical analysis and demonstrate the performance of AAZA in this section. The network topology for simulation is given in Fig. 5.11, which is used in the previous chapter. The black node denotes the streaming server (e.g., AAZA server) that realizes different channel zapping acceleration schemes. In the simulation, an MPEG-4 video clip is segmented into packets and multicast to STBs upon request. These packets can be reverted to the video format and displayed with dedicated program [9]. The largest GOP size of the video is set to 1,000 ms. The time for each STBs to issue zapping requests is randomly distributed over the duration of the video clip.

5.5.1 Visual Effect

We examine the visual effect observed by the STB under AAZA and MICC [11]. Figure 5.12a displays the first pictures received after the zapping request under the two schemes; Fig. 5.12b shows the pictures seen at the STB just before its migration from the sCH to mCH. For both of the figures, the picture on the left is yielded with AAZA and the one on the right with MICC. Pictures in Fig. 5.12a are retrieved from

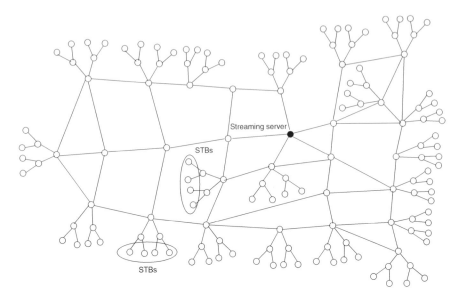

Fig. 5.11   Simulation topology

an STB that is 3 hops away from the streaming server, while the STB observing pictures in Fig. 5.12b is 6 hops away from the streaming server. The visual effect of AAZA is obviously better than MICC, which can be identified by observing the man's eyeballs in the pictures. Moreover, Fig. 5.12b shows the MICC scheme could incur the picture inconsistency when the STB is migrating from the auxiliary stream to the main stream. This is because the AAZA sCH for fast zapping is a full-quality stream media while the auxiliary stream of MICC just contains I-frames. With AAZA, receivers can always see the full-quality pictures of the video even in the duration of channel zapping.

5.5.2   First I-Frame Delay

In this section, we examine the FID performance of AAZA with different STBs. We focus on FID as it is the main contributor of zapping time in IPTV. The impact of the population of STBs N on the FID by AAZA is shown in Fig. 5.13, with a fixed value of $T = 200$ ms. As shown in Fig. 5.13a, the average FID is around 100 ms for different values of N. This is because the zapping requests will definitely fall into an interval sized T to bound the FID, which has been proved in Theorem 5.1. Moreover, the sending time of these requests are randomly distributed over the duration of the video playing out. Thus, the average value of the FID should be approaching to $\frac{T}{2}$. The cumulative distribution functions (CDFs) of the FID illustrated in Fig. 5.13b support this view. The results from the scenarios

a

First pictures observed

b

Pictures observed just before migration

Fig. 5.12   Visual effects (*left*: AAZA, *right*: MICC)

where STBs are sparsely distributed ($N = 16$) and densely distributed ($N = 80$) are presented. We can see that STBs with FIDs less than 100 ms account for about 50% for both cases, and basically, the FIDs are evenly distributed over T.

5.5.3   Optimal Subchannel Data Rate

The optimal sCH data rate R^* has been derived. We here set the ratio of main channel data rate and sCH data rate as different values and see if $R^* = 2r$ indeed makes the sCH transmit the least redundant traffic.

The average amount of data transmitted over sCHs with different values of $\Delta = \frac{R}{r} - 1$ are illustrated in Fig. 5.14a. It is clear that the sCHs transmit the fewest amount of data when $\Delta = 1$, which means $R = R^* = 2r$. In the simulation, the "theoretical" value is the amount of data transmitted over all activated sCHs from the times they are turned on to the moments they are turned off. In practical, it may happen that the STB joins in a sCH that has been turned on for a while; the "experimental" value is the amount of sCH data actually received by STBs. As the zapping requests reach

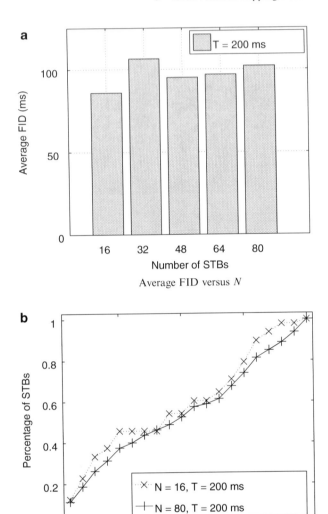

Fig. 5.13 The impact of N on FID by AAZA

the AAZA server at different times, sCHs to be activated are different for the sparse and dense cases. Moreover, the lifetimes of sCHs are different according to their sCH indices. This is why the theoretical (experimental) values are different in the cases of $N = 16$ and $N = 80$.

The distributions of the amount of data transmitted over a sCH are illustrated in Fig. 5.14b for $N = 80$. When $R = R^*$, the percentage of sCHs with light data

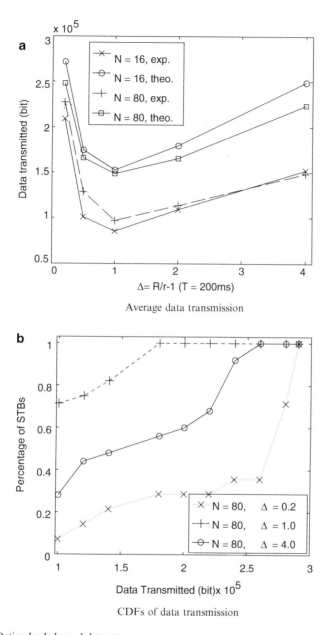

Fig. 5.14 Optimal subchannel data rate

transmitting is higher than that when R takes other values. It shows that all sCHs transmit less than 2×10^5 bits with $R = R^*$, while some sCHs need to transmit about 3×10^5 bits if R takes other values.

5.5.4   Signaling Delay Effect

In MAZA, the signaling delay can make the zapping request miss the I-frame expected, thus affect the FID performance. We examine the signaling delay effect by configuring extra traveling time W for zapping requests.

The average FIDs perceived in MAZA with different values of W are shown in Fig. 5.15a, for $N = 80$. As MAZA is also an implementation of the TSS-based service model, the average FIDs under MAZA and under AAZA are the same when $W = 0$. However, the FID performance in MAZA is significantly affected even with a small value of W. The curves representing the average FID first increase dramatically, then decrease, and gradually approach to some level. Both T and W can impact the FID performance of MAZA. If T is large, it is more probably that the zapping request will not miss the expected I-frame even with the extra traveling time, and the signaling delay could even luckily further mitigate the FID; however, if T is small, it is easier for the zapping request to fall out of the interval sized T with the extra delay, and the STB has to wait for the I-frame in the next GOP. Similarly, a larger W will make the I-frame slip away more easily. That is the reason the longest average FID occurs with $T = 100\,$ms and $W = 50\,$ms, and it becomes smaller when T increases. When T becomes much larger than W, the impact from the signaling effect will be wakened as most of the zapping requests cannot miss the expected I-frame even with extra traveling time, and the average FID will converge to around $\frac{T}{2}$; however, large T will negatively impact the FID performance of IPTV systems.

Many lucky zapping requests can be found when T becomes larger. The extra delay for these requests can reduce the FID by T at most; however, if an I-frame is missed, the next I-frame could be up to XT time units away. The negative side of the signaling delay effect is still dominant. Figure 5.15b shows this fact, where the number of STBs with FIDs beyond $T = 300\,$ms account for about 30–40% if $W \neq 0$.

5.5.5   Control Messages Overhead

We count the total number of control messages all relay nodes have seen within $10\,$s since the video starts to play under MAZA and AAZA. The total number of received messages by all relay nodes versus different STB populations is depicted in Fig. 5.16. Most of the messages for MAZA are from meta-channel which broadcasts to STBs each time an I-frame appears over the main channel. There are 264 I-frames identified by the MAZA server within $10\,$s in the simulation. Because of its broadcasting nature, the total number of control messages over the network will grow linearly with the number of STBs in the network. In contrast, AAZA only needs the leave-and-join operation for zapping, and the seamless migration from the sCH to mCH is achieved by modifying the destinations list. Thus, the number of control messages is negligible compared with that in MAZA. Although the control messages may not consume much bandwidth resources, they incur more chances of

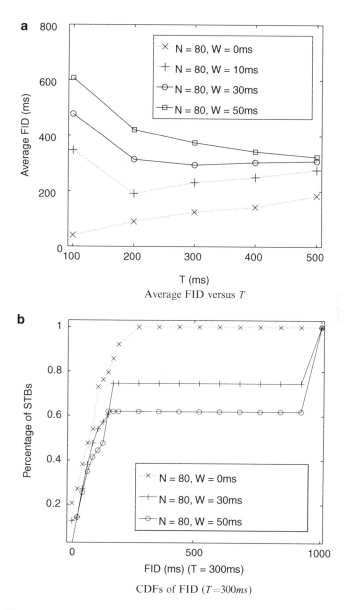

Fig. 5.15 FID performance with signaling delay effect

system errors. For example, if an STB missed one or two meta-channel messages when it wants to zap channel, the FID could be more than T. Thus, AAZA provides a higher robustness to the IPTV systems compared with MAZA.

Fig. 5.16 Control messages overhead

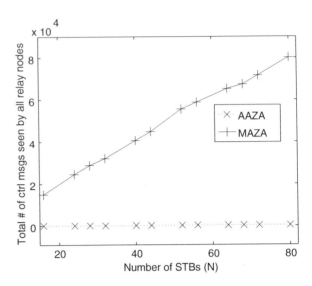

5.5.6   Data Forwarding Overhead

This section examines the bandwidth overhead incurred by data forwarding under AAZA and MAZA. We count the number of packets received by each node for 10 s since the video starts to play. The average numbers and CDFs of packets receptions within 10 s are illustrated in Fig. 5.17. The explicit addressing in AOM causes an extra bandwidth consumption of AAZA compared with MAZA, when the number of STBs are densely populated; however, the difference is small. AOM encodes destination addresses list into the packet to facilitate the data forwarding. If the number of addresses is beyond the capacity of the list, multiple packets have to be generated to cover all the receivers. In the case of $N = 16$, the list has enough space to contain all destination addresses, thus, the bandwidth consumptions of AAZA and MAZA are the same. When $N = 80$, AOM has to generate some redundant traffic to cover all destinations; therefore, the bandwidth overhead of AAZA is higher than that of MAZA, as illustrated in Fig. 5.17a. However, as AOM [15, 16] leverages some techniques to constrain the extra bandwidth cost, the difference of bandwidth consumptions between AAZA and MAZA is insignificant, as illustrated in Fig. 5.17b.

5.6   Summary

In this chapter, we have applied the proposed AOM scheme to the IPTV systems to reduce the channel zapping time, which is an important QoE metric. We have briefly introduced the TSS-based service model and presented a systematical analysis of

Fig. 5.17 Data forwarding overhead

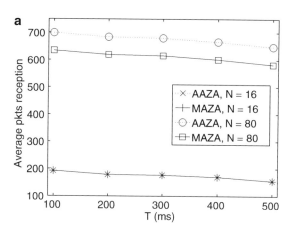

Average packet reception

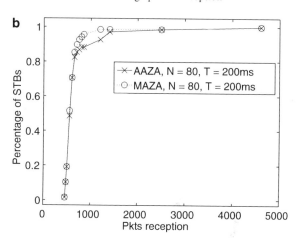

CDFs of the packets reception per node

the TSS model. We have shown that there exists an optimal subchannel data rate which minimizes the redundant traffic transmitted over subchannels; moreover, we have revealed the *start-up effect*, where the subchannel data rate and the time shift between successive sub-channels can impact the behavior of the TSS-based model. With a convenient solution to the start-up effect proposed, we have provided a rigorous proof that the bounded FID can be always guaranteed by the TSS-based model. Furthermore, we have proposed the AAZA scheme, which implements the TSS-based model with the facilitation of AOM. AAZA achieves the seamless subscriber migration from the subchannel to the main channel, without any control message exchange over the network. The sub-channel selection is independent of the zapping request signaling delay, resulting in improved robustness and reduced messaging overhead of AAZA. We have implemented AAZA in ns-2 and multicast an MPEG-4 video stream over a practical network topology. Extensive simulation results have been presented to demonstrate the validity of our analysis and AAZA scheme.

References

1. Bejerano Y, Koppol PV (2009) Improving zapping response time for IPTV. In: Proc. IEEE INFOCOM, Mar 2009, pp 1971–1979
2. Boyce J, Tourapis A (2005) Patent (wo/2005/112465): method and apparatus enabling fast channel change for dsl system (assignee: Thomson), Nov 2005
3. Boyce JM, Tourapis AM (2005) Fast efficient channel change. In: Proc. International Conference on Consumer Electronics, Jan 2005
4. Cisco systems, Optimizing video transport in your IP triple play network. http://www.cisco.com/en/US/prod/collateral/routers/ps368/prod_white_paper0900aecd80478c12.html
5. Cohen N (2006) USPTO patent application 20060143669: fast channel switching for digital TV (assignee: Bitband technologies (filed)), Jun 2006
6. Dacosta B (2006) USPTO patent application 20060230176: methods and apparatus for decreasing streaming latencies for IPTV (assignee: Sony (filed)), Oct 2006
7. Farmer J (2006) USPTO patent application 20060075428: minimizing channel change time for IP video (assignee: Wave7 optics (filed)), Apr 2006
8. Hong CK, Lee CY, Lee KY (2010) Reducing channel zapping time in IPTV based on user's channel selection behaviors. IEEE Trans Broadcasting PP:1–10
9. Hwang W, Ke C, Shieh C, Ziviani A (2008) An evaluation framework for more realistic simulations of MPEG video transmission. J Inf Sci Eng 24:425–440
10. Jun Y, Cho C, Han I, Lee H (2004) Improvement of channel zapping time in IPTV services using the adjacent groups join-leave method. In: Proc. 6th International Conference on Advanced Communication Technology, 2004
11. Kalyanaraman S, Gerber A, Banodkar D, Ramakrishnan KK, Spatscheck O (2008) Multicast instant channel change in IPTV systems. In: Proc. 3rd International Conference on Communication Systems Software and Middleware, 2008, pp 370–379
12. Pelt M (2005) USPTO patent application 20050265374: broadband telecommunication system and method used therein to reduce the latency of channel switching by a multimedia receiver (assignee: Alcatel (filed)), Dec 2005
13. Shaikh A, Wang J, Yates J, Zhang Y, Mahimkar A, Ge Z, Zhao Q (2009) Modeling channel popularity dynamics in a large IPTV system. In: Proc. ACM SIGMETRICS, 2009, pp 275–286
14. The network simulator – ns-2. http://www.isi.edu/nsnam/ns
15. Tian X, Cheng Y, Liu B (2009) Design of a scalable multicast scheme with an application-network cross-layer approach. IEEE Trans Multimedia 11:1160–1169
16. Tian X, Cheng Y, Shen X (2010) DOM: a scalable multicast protocol for next-generation Internet. IEEE Network 24:45–51
17. Tian X, Cheng Y, Liu B (2011) A generic application-oriented networking (GAON) simulation framework for next-generation internet. In: Proc. IEEE ICC
18. De Vleeschauwer D, Degrande N, Laevens K, Sharpe R (2008) Increasing the user perceived quality for IPTV services. IEEE Commun Mag 46:94–100
19. Zhang C, Wu D, Ross KW (2011) Unraveling the BitTorrent ecosystem. IEEE Trans Parallel Distr Syst 22:1164–1177

Chapter 6
Generic AON (GAON) Simulation Framework

Abstract The AOM multicast mechanism proposed in previous chapters is realized with an AON approach, where network routers are enhanced with application-oriented intelligence. In fact, AON is also one of the mainstream ideas to design the next-generation Internet. However, there is no systematic study on what intelligence should be incorporated into the router and what the fundamental benefit of the AON approach could bring. Resorting to simulation tools is an important way to fully understand and address these issues. The network simulator ns-2 is the most accepted simulation tool in the networking community, but instructions on how to modify the ns-2-wired node structure to implement developer-defined processing are still limited. This chapter presents a generic application-oriented networking (GAON) simulation framework compatible with ns-2 to facilitate the AON research. The practical ns-2 simulation techniques introduced in this chapter could also be useful for general ns-2 developers. The user case diagram and the extended ns-2 node structure are presented in Sect. 6.1. In Sect. 6.2, key OTCL and C++ interfaces of GAON framework are described, which also reveals the fundamental principles of ns-2. In Sect. 6.3, we present the implementation of AOM under the GAON framework to demonstrate the GAON development process. The chapter is concluded in Sect. 6.4.

6.1 GAON: An Overview[1]

6.1.1 GAON Framework

We use the unified modeling language (UML) [7] user case diagram in Fig. 6.1 to illustrate the functionalities of GAON and how GAON components interact between

---

[1] ©[2011] IEEE. Portions reprinted, with permission, from [6].

X. Tian and Y. Cheng, *Scalable Multicasting over Next-Generation Internet: Design, Analysis and Applications*, DOI 10.1007/978-1-4614-0152-0_6, © Springer Science+Business Media New York 2013

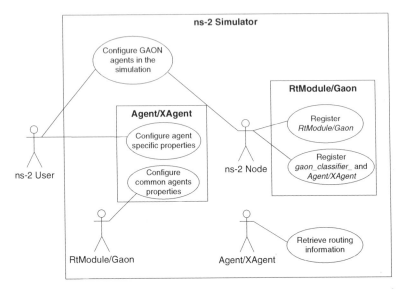

Fig. 6.1 User case diagram of GAON

each other. The rectangle represents the boundary of a functional component, and the ellipsis denotes the functions the component provides to users that are represented as matchmen.

As in Fig. 6.1, the ns-2 *Simulator* provides the ns-2 user an interface to configure GAON agents, through which the user can load/unload specific GAON agents on each node. The configuration information will be received by the node creation component *Node*, which then creates network nodes as user configured. For example, *Node* equips each instance node with an *Agent/XAgent* in Fig. 6.1. This is done by utilizing the functionalities provided by *RtModule/Gaon*, where the *Agent/XAgent* and other GAON components are properly installed in an ns-2 node. The ns-2 *Simulator* also provides an interface to *Agent/XAgent* for retrieving the ns-2 built-in routing information so that the agent can correctly forward the packet after processing. In addition, *Agent/XAgent* has interfaces to the ns-2 user and *RtModule/Gaon* for agent-specific and common agents properties configuration, respectively.

6.1.2 GAON Node Model

The GAON node is created as in Fig. 6.2, where the network-layer capability is enhanced with the application-oriented intelligence incarnated by GAON agents (i.e., *xagent_, yagent_*). GAON router classifies the incoming network traffic as *normal traffic* and the *GAON traffic* that needs application-oriented processing.

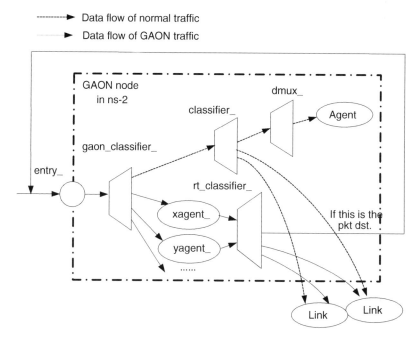

Fig. 6.2 Structure of GAON node in ns-2

The former will be directly forwarded against the unicast routing table, and the latter will be first dispatched to corresponding GAON agents for specific processing and then forwarded (recall Fig. 2.11 in Sect. 2.4).

Specifically, a flag variable *gaon_mod_valid* can be defined in *struct hdr_cmn* of ns-2 file "Packet.h" to identify the type of the packet. When a packet arrives at a GAON node, it is first classified by the *gaon_classifier_*. If the packet has *gaon_mod_valid = 0*, meaning that it is *normal*, the packet will be passed to the regular ns-2 *classifier_*, which further delivers the packet to upper layer agent (if this node is the packet's destination) via port classifier *dmux_* or to the corresponding link object for the next-hop node. If the flag indicates that the packet needs GAON agent's processing, i.e., *gaon_mod_valid > 0*, the packet will be passed to the corresponding agent based on the value of the flag variable. After specific processing, the GAON packet goes through the *rt_classifier_*, which obtains the corresponding link object and forwards the packet to the next-hop node based on the packet's destination address. If *rt_classifier_* finds that the current node is the destination of the GAON packet, it will set *gaon_mod_valid = 0* and send the packet back to the entrance of the node again. Then the packet will follow the data flow of normal traffic and finally reach the upper layer agent.

The next section focuses on describing the interfaces between GAON and regular ns-2, where the following issues corresponding to Fig. 6.1 will be addressed:

- How to implement the user-Simulator interface to load/unload the specific GAON agents on the ns-2 node.
- How to implement the interface between *Node* and *RtModule/Gaon*.
- How GAON nodes can be created with the structure as in Fig. 6.2.
- How to develop a unified interface for GAON agents to retrieve the routing table built in ns-2.

6.2 GAON Interfaces to ns-2[2]

6.2.1 Scenario Control Interface

GAON provides a scenario control interface for users to flexibly load/unload certain GAON agents on a specific node. Program 1 presents a sample simulation script, where the user first creates a *Simulator* object and then specifies which GAON agents should be loaded on the nodes to be created. The names of these agents are recorded in the list *gaonagent_list_*, which is added as an instance variable of class *Simulator* in the OTcl domain. For example, the codes in Program 1 mean that the user wants nodes *n(0)* and *n(1)* to load objects of classes *Agent/XAgent* and *Agent/YAgent*, and node *n(2)* to be a regular ns-2 node without any GAON component. When trying to load/unload other nodes with different GAON agents, the user only needs to manipulate the *gaonagent_list_* and invoke the procedure *gaon_on*. The symbol "\" is the line breaker in Tcl language [5].

```
set ns [new Simulator]                                    1
                                                          2
[$ns set gaonagent_list_] lappend XAgent,YAgent           3
                                                          4
$ns gaon_on 1                                             5
set n(0) [$ns node]                                       6
set n(1) [$ns node]                                       7
$ns gaon_on 0                                             8
set n(2) [$ns node]                                       9
. . . . . .                                               10
```

Program 1. Sample simulation tcl script

The scenario control interface is implemented in the OTcl domain of ns-2. The instance variable *gaonagent_list_* is declared when the object of *Simulator* is initiated as illustrated in line 4 of Program 2. When the instance procedure *gaon_on* is invoked to enable the GAON routing module *RtModule/Gaon*, the procedure will

---

[2]©[2011] IEEE. Portions reprinted, with permission, from [6].

add the string "Gaon" to the class *Node*'s built-in instance variable *module_list_* by invoking the instance procedure *enable-module* of class *Node* as shown in line 9. The list *module_list_* is declared as in line 17 and stores all routing modules that will be registered to the node. Similarly, instance procedure *disable-module* of *Node* can delete the "Gaon" from *module_list_*; thus, the routing module *RtModule/Gaon* will not be registered to the node, and the node to be created will be a regular unicast node.

```
//~/ns/tcl/lib/ns-lib.tcl                                         1
Simulator instproc init args {                                    2
    ......                                                        3
    $self instvar gaonagent_list_                                 4
}                                                                 5
                                                                 6
Simulator instproc gaon_on { enable_ } {                         7
    if ( enable_ == 1 ) {                                        8
        Node enable-module Gaon                                   9
    } else {                                                     10
        Node disable-module Gaon                                 11
    }                                                            12
}                                                                13
                                                                14
//~/ns/tcl/lib/ns-node.tcl                                       15
    ......                                                       16
    Node set module_list_ {Base}                                 17
```
Program 2. Scenario control interface in Simulator

6.2.2   Installing GAON Components in an ns-2 Node

The *RtModule/Gaon* object is registered to the node through executing its own instance procedure *register* as in Program 3. In *register* procedure, the GAON agent classifier and GAON agents are installed in the ns-2 node as shown in Fig. 6.2. The expression "RtModule/Gaon" means that class *RtModule/Gaon* is derived from its parent class *RtModule* in OTcl domain, where the class *RtModule* is purposely designed to facilitate the management and configuration of classifiers in ns-2 nodes.

```
//~/ns/tcl/lib/ns-rtmodule.tcl                                   1
RtModule/Gaon instproc register { node } {                       2
  $self next $node                                               3
  $self instvar classifier_                                      4
  $self set classifier_ [$node entry]                            5
  $node set gaon_classifier_ [new Classifier/GAON]               6
  $node insert-entry $self \                                     7
                    [$node set gaon_classifier_] 0               8
                                                                 9
  set lstlen [llength [[Simulator instance] \                   10
                    set gaonagent_list_]]                       11
                                                                12
```

```
for {set i 0} {$i < $ lstlen} {incr i} {          13
   set agent_item [lindex \                        14
   [[Simulator instance] set gaonagent_list_] $i]  15
                                                    16
      if { $agent_item == " XAgent"} {             17
         $node set xagent_ \                        18
                [new Agent/$agent_item]            19
         [$node set gaon_classifier_] install \    20
                 1 [$node set xagent_]             21
         $node attach [$node set xagent_]          22
                                                    23
      } else if { $agent_item == "YAgent" }{       24
         $node set yagent_ \                        25
                [new Agent/$agent_item]            26
         [$node set gaon_classifier_] install \    27
                 2 [$node set yagent_]             28
         $node attach [$node set yagent_]          29
      }                                             30
      . . . . . .                                   31
   # add your new agent here                        32
   }                                                33
}                                                   34
```

Program 3. Installing GAON components in a node

In Program 3, the first step is to register the *RtModule/Gaon* itself to the node being created and set up the instance variable *classifier_* of *RtModule* to be the first object connecting from the entry of the node (lines 3–5). As of now, the regular *classifier_* of the ns-2 node is established. When executing line 6, the GAON agent classifier *gaon_classifier_* is created, in which there is a key variable *slot_*. The variable *slot_* is an array of pointers, which stores pointers to downstream NsObjects matching different criteria. To form the structure as illustrated in Fig. 6.2, the *Node* object node first invokes the procedure *insert-entry*, inserting the *gaon_classifier_* as the head classifier connecting from the node entry, and installs the existing head classifier *classifier_* in the slot 0 of the *gaon_classifier_* (lines 7–8). Then, for each agent name in the *gaonagent_list_* configured through the scenario control interface, if the agent name is "XAgent," an object of class *Agent/XAgent* is created and installed in the slot 1 of *gaon_classifier_*. Similarly, the object of *Agent/YAgent* is installed in the slot 2 of *gaon_classifier_*. The instance procedure *install* of the object node is used to do that (lines 20–21, 27–28). The user could add new GAON agent by installing it in other slots of *gaon_classifier_*. In this way, the GAON components are installed in the ns-2 node just as illustrated in Fig. 6.2.

6.2.3   GAON Agent Classifier

The classifier in ns-2 is a packet forwarder, which forwards the incoming packet to other ns-2 objects based on certain criteria. Our GAON agent classifier is derived from ns-2 class *Classifier*. Program 4 gives the implementation of *GAONClassifier* (corresponding to OTcl class *Classifier/GAON* in Program 3), where only the function *find(p)* is overridden to classify the incoming packet according to the value of *gaon_mod_valid* as shown in lines 9–14. As described in Sect. 6.1.2, the *slot_* of *gaon_classifier_* is properly populated when registering *RtModule/Gaon*, where *slot_[0]* stores the *classifier_*, *slot_[1]* stores the *xagent_*, and *slot_[2]* stores the *yagent_*.

```
// ~/ns/gaon_module/classifier_gaon.h                    1
class GAONClassifier : public Classifier {              2
    protected :                                          3
        NsObject *find(Packet *);                        4
};                                                       5
                                                         6
// ~/ns/gaon_module/classifier_gaon.cc                   7
NsObject *GAONClassifier :: find(Packet *const p) {     8
    hdr_cmn *cmh = hdr_cmn :: access(p);                 9
    if (cmh->gaon_mod_valid == '0') {                    10
        return slot_[0];                                 11
    }                                                    12
    else if (cmh->gaon_mod_valid == '1') {               13
        return slot_[1];                                 14
    }                                                    15
    ......                                               16
}                                                        17
```

Program 4. Implementation of GAONClassifier

6.2.4   Interface to ns-2 Routing Table

We provide a function *RtClassifier()* to fulfil the task of *rt_classifier_* in Fig. 6.2, which returns the NsObject leading to the next-hop node given the forwarding node address and the packet's destination address. This design is simpler and more efficient compared with creating a new classifier object. It is because there has been a class *RouteLogic* in ns-2, which computes the optimal routes and creates the routing table in the OTcl domain; however, most developers prefer realizing their agents in C++ domain. Thus, the only thing needed to be done is implementing a friendly interface to access the routing table from C++ domain. This interface is implemented in Program 5, where the header file "tclcl.h" is included so that the gateway *Tcl::instance()* from C++ domain to the entire OTcl domain defined in ns-2 is accessible. Lines 8–12 show how to get the reference of the ns-2 built-in routing table and retrieve the next-hop node address based on the forwarding node address and the packet's destination address. Lines 14–23 show how to get the link head object leading to the next-hop node. The resulted object is named as in OTcl domain, and stored in *nhop_link*. Codes in lines 24–25 retrieve the shadow object in C++ domain corresponding to the resulted OTcl object. By directly retrieving the ns-2 built-in routing table, the latest routing information is always available for each GAON agent. When the dynamic link failure is considered in the simulation scenario, the routing table will be automatically updated.

The function *RtClassifier()* is invoked only if the forwarding node is not the packet's destination; if not, the GAON agent only needs to invoke the member function *send()* of its parent class *Agent* to send the packet back to the entry of the node, as illustrated in Fig. 6.2.

```
// ~/ns/gaon_module/RouteClassifier.h                    1
#include <tclcl.h>                                       2
                                                         3
NsObject *RtClassifier(Packet *p, int CurAddr) {         4
    hdr_ip *iph = hdr_ip::access(p);                     5
    int Dst = iph->daddr();                              6
    Tcl& tcl = Tcl::instance();                          7
    char next_hop_id[10];                                8
    tcl.evalf("[[Simulator instance] \                   9
                get-routelogic] \                        10
                lookup %d %d'', CurAddr, Dst);           11
    sprintf(next_hop_id, "%s", tcl.result());            12
                                                         13
    int nhop_id = atoi(next_hop_id);                     14
    char nhop_link[10];                                  15
    tcl.evalf("[[Simulator instance] \                   16
                link \                                   17
                [[Simulator instance] \                  18
                    get-node-by-id %d] \                 19
                [[Simulator instance] \                  20
                    get-node-by-id %d]] head", \         21
                CurAddr, nhop_id);                       22
    sprintf(nhop_link, "%s", tcl.result());              23
    NsObject* nxhop =                                    24
        (NsObject*)TclObject::lookup(nhop_link);         25
    return nxhop;                                        26
}                                                        27
```

Program 5. Definition of function RtClassifier

6.3   Implementation of AOM Under GAON

This section describes how to implement AOM in ns-2 under the GAON framework. As presented in the previous section, GAON framework takes the responsibility of interconnecting the developer-defined agents and ns-2 utilities, so the AOM protocol can be largely implemented using C++. The following files will be created in a

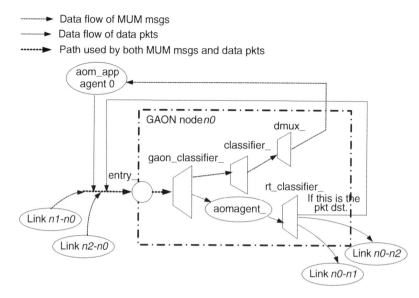

Fig. 6.3 AOM application-layer agent with GAON node

new directory $\tilde{/}ns/gaon\_module/aom\_protocol$, which realize the enhancement of
AOM intelligence into the network routers:

- **bloom_filter.h/.cc**. The file where a series of hash functions are declared/defined.
- **aomagent.h**. The header file where the AOM network-layer data structures and
  functions are declared.
- **aomagent.cc**. This file binds the AOM network-layer agent to OTCL interface
  and implements AOM protocol functionalities of SDR and TBR.

The following files will be created in the directory $\tilde{/}ns/apps$, which realize the
AOM application-layer agent:

- **aom_app.h**. The header file where the AOM application-layer data structures and
  functions are declared.
- **aom_app.cc**. This file binds the AOM application-layer agent to OTCL interface
  and implements AOM protocol functionalities of SRC and RDR.

The relationship between network-layer AOM agent and application-layer AOM
agent is illustrated in Fig. 6.3. The *aomagent_* realizes the functionalities of
SDR/TBR in the network, while the *aom_app* realizes the functionalities in SRC
and RDR. Consider a simple topology with three nodes $n0$, $n1$, and $n2$, and two
links originated from $n0$, the *aom_app* agents attached to $n1$ and $n2$ sent two MUM
messages to $n0$. When the MUM messages arrive at the entry point of $n0$, they will
be passed to the *aom_app* agent attached to $n0$, and the *aom_app* agent will perform
the SRC processing introduced in Sect. 3.3.3. After the AOM packet shim header is
inserted into the multicasting data packet, the packet will be dispatched to receivers
($n1$ and $n2$) along the flow of data packets as shown in Fig. 6.3.

6.3.1 Packet Header

The AOM protocol defines a new packet type which will carry the shim header and relevant control data structures. The new packet type definition is put in the file *aom_app.h* as the following:

```
// ~/ns/apps/aom_app.h                                                   1
#include <packet.h> struct hdr_aom_app {                                 2
  nsaddr_t src_addr; // The address of the SRC node                      3
  char mum_msg_rdr[20]; // The prefix the RDR which sends                4
                        // the MUM message                               5
  bloom_filter* mum_GRP_BF; // The GRP_BF in the MUM msg                  6
  IntVec* mum_srt_path;   // BGP view implementation, source             7
                          // routing path to reach src_addr              8
  IntVec* mum_fast_join; // The grp ids for fast join                    9
  int aom_msg_type; // The type of the AOM message,                     10
                    // joining, leaving or data forwarding              11
  nsaddr_t reverse_interface; // The logical interface index            12
                              // installed at TBR                       13
  bloom_filter* DST_BF_pkt; // DST_BF in the data pkt                   14
  ......                                                                15
                                                                        16
  static int offset_;                                                   17
  inline int& offset() { return offset_; }                              18
  inline static hdr_aom_app* access(Packet* p){                         19
  return (hdr_aom_app*) p->access(offset_);                             20
  }                                                                     21
};                                                                      22
```

Program 6. AOM packet header definition

In Program 6, **nsaddr_t** is the network address type of ns-2, **bloom_filter** type will be described in Sect. 6.3.3, and **IntVec** is the type representing a vector of integers. The AOM header contains data structures for both MUM messages and data packets. The message type is specified by *aom_msg_type*. In order to join in an SRC, the RDR first has to know the *src_addr*. The RDR also specifies its prefix for forwarding states installation. The GRP_BF in the MUM message is to reduce the bandwidth cost as described in Sect. 3.3.2. The MUM message should be supported by the source routing in the asymmetric routing environment for which the routing path is configured in the *mum_srt_path*. The *reverse_interface* is for supporting the incremental deployment. After the SRC processing, the destinations of a group will be encoded in the DST_BF to facilitate downstream forwarding. Lines 18–21 define the method to have access to the AOM header.

The AOM header needs to be bound to OTCL interface, so that it can be accessed through OTCL domain. The binding can be done as shown in Program 7. In fact, the binding operation is a routine in ns-2 development. Every new header defined in the C++ domain can be bound to the OTCL interface using the similar codes as shown in Program 7. We only need to change the corresponding header name. The packet header structures and access principles are described in detail in [4].

```
// ~/ns/apps/aom_app.cc                                    1
int hdr_aom_app::offset_;                                  2
static class AOMAppHeaderClass : public                    3
PacketHeaderClass {                                        4
public:                                                    5
    AOMAppHeaderClass() :                                  6
    PacketHeaderClass("PacketHeader/AOMApp",               7
                      sizeof(hdr_aom_app)) {               8
        bind_offset(&hdr_aom_app::offset_);                9
    }                                                      10
} class_aomapphdr;                                         11
```

Program 7. AOM packet header bound to OTcl interface.

6.3.2   Application-Layer AOM Agent

The application-layer AOM agent performs different processing based on what kind of node it is attached to. If the agent is attached to an SDR, it plays the role of an SRC, which generates multicasting traffic and prepares DST_BFs for multicasting packets. If the agent is attached to RDR, it is responsible for sending control messages. Program 8 shows the declaration of application-layer AOM agent.

```
// ~/ns/apps/aom_app.h                                     1
class AOMAppAgent;                                          2
class TimerMainCh : public TimerHandler {                  3
public:                                                    4
    TimerMainCh(AOMAppAgent *aomappagt){                   5
    aomappagent_ = aomappagt;                              6
    };                                                     7
    virtual void expire( Event *e);                        8
protected:                                                 9
    double next_pkttime_;                                  10
    AOMAppAgent *aomappagent_;                             11
};                                                         12
    . . . . . .                                            13
class AOMAppAgent : public Agent { public:                 14
    AOMAppAgent(): mch_timer_(this);                       15
                                                           16
    IntSet local_chnellst_; // Local channel list at SRC   17
    MUMTable mum_table_; // MUM Table at SRC               18
    McastDstsCache cached_shmhdrs_; // Dsts cache table or AOM  19
                          // Destinations Cache (ADC)       20
    int src_bfsz_; // GRP_BF size                          21
    int mum_entry_vol_; // GRP_BF capacity                 22
    int tbr_bfsz_; // IRDR_BF size                         23
    int shim_hdr_vol_; // Shim header capacity             24
    Mpeg *trace_; // Structure holding MPEG frames         25
    void SRCProcMumMsg(); // MUM message processing at SRC 26
    void SRCProcDstsCache(); // Update the dsts. cache      27
```

```
void SendFrameMainCh (); // Send  data  pkts  over  mCH        28
void AddAOMHdr(); // Insert  AOM  header  to  the  data  pkts   29
     ......                                                     30
virtual int command(int argc, const char*const* argv);         31
virtual void recv(Packet*, Handler*);                          32
protected :                                                     33
     TimerMainCh mch_timer_;                                    34
}                                                              35
```

Program 8. Declaration of application-layer AOM agent

We first consider the SRC case. In order to generate multimedia stream, an MPEG-4 video clip is segmented into packets in the SRC [2], and each packet is dispatched to receivers after a period of time. The data structure in line 27 is for holding these packets. Each packet will be sent out after a period of waiting time, which is counted by the time $mch\_time\_$ (line 35). This time cross-references with the object of *AOMAppAgent*. The constructor of class *AOMAppAgent* constructs its variable $mch\_timer\_$ with the pointer *this* to the *TimerMainCh* object as shown in line 16. In line 6, the constructor of class *TimerMainCh* stores the input pointer in variable $aomappagent\_$; thus, the pointer points back to the object of class *AOMAppAgent*. In this way, each *TimerMainCh* and *AOMAppAgent* object is associated with each other. The *TimerHandler* is an ns-2-defined class representing a timer. In the following description, some details of Program 8 will be shown in Program 9 and 10 for convenient demonstration.

```
// ~/ns/apps/aom_app.cc                                         1
static class AOMAppClass : public TclClass { public :          2
    AOMAppClass () : TclClass ("Agent/AOMApp") {}               3
    TclObject* create(int , const char*const *) {               4
        return (new AOMAppAgent());                             5
    }                                                           6
} class_aomapp; // Bind  AOMApp  agent  to  OTcl  interface void 7
TimerMainCh :: expire (Event* e){                               8
    aomappagent_->SendFrameMainCh ();                           9
} void AOMAppAgent:: SendFrameMainCh (){                        10
    // Retrieve  pkts  from  trace_                             11
    ......                                                      12
    AddAOMHdr();                                                13
    // Set  next  timer  expiration  interval                   14
    mch_timer_.next_pkttime_ = time_to_wait;                    15
    mch_timer_.resched (time_to_wait);                          16
}                                                              17
```

Program 9. Timer and AOMApp agent

The timer expires after waiting $next\_pkttime\_$ seconds. The expiration event will trigger the function *SendFrameMainCh()* to send the next packet out. This procedure is realized in Program 9. The first task is to associate the name *Agent/AOMApp* in the OTCL domain with the object created by invoking constructor *AOMAppAgent()* (lines 2–10). When the packet is delivered, the function *AddAOMHdr()* is invoked to insert the DST_BF to the multicasting packet.

Lines 18–21 of Program 8 define necessary data structures for the agent to prepare the DST_BF. As shown in lines 5–8 of Program 10, these structures are leveraged by functions *SRCProcMumMsg()* and *SRCProcDstsCache()* to aggregate the MUM messages and update the destination cache, respectively. The information retrieved from the AOMApp header (lines 3–4 of Program 10) specifies the AOM message type. The SRC performs corresponding process described in Sect. 3.3.3.

```
// ~/ ns / apps / aom_app . cc                                            1
void AOMAppAgent :: recv ( Packet ∗  pkt ,  Handler ∗ ){             2
    hdr_ip∗ hdrip = hdr_ip :: access ( pkt );                        3
    hdr_aom_app∗ aomhdr = hdr_aom_app :: access ( pkt );             4
    if ( aomhdr −>aom_msg_type  == 0){  // If received a join msg  5
        SRCProcMumMsg ( );    // MUM message processing at SRC      6
        SRCProcDstsCache ( );   // Update the dsts. cache           7
    }                                                                8
    . . . . . .  // Process other types of msgs                      9
}                                                                   10
```
Program 10. AOM message processing

For the case where the AOMApp agent is attached to an RDR, the agent should be able to send control messages, i.e., joining, leaving, etc. Program 11 shows such an implementation. When a command *send_join* is received from the OTCL domain, the AOMApp agent creates a packet for joining message. The *aon_mod_valid* property of the packet is set so that the packet can be passed to the AOM agent in the network layer for some processing, e.g., source routing (lines 4–9). The message type is correspondingly configured to make sure the network-layer agent can correctly process the message.

Recall the attributes in lines 22–25 of Program 8, which are for holding the bloom filter properties. They will be configured by users through the OTCL interface, as shown in lines 20–27 of Program 11.

```
// ~/ ns / apps / aom_app . cc                                             1
int AOMAppAgent :: command ( int argc ,  const char ∗const ∗  argv ){  2
    if ( argc  == 2) {                                               3
        // Send join msg to SRC                                      4
        if ( strcmp ( argv [ 1 ] , ”send_join”) == 0) {             5
            Packet ∗ pkt = allocpkt ( );                            6
            hdr_cmn ∗cmh = hdr_cmn :: access ( pkt );              7
            cmh−>aon_mod_valid = ’1’;    // Go to AOM agent        8
            hdr_aom_app∗ aomhdr = hdr_aom_app :: access ( pkt );   9
            aomhdr−>aom_msg_type = 0;  // It is a join msg         10
            // Assign values to data structures in the header      11
            // e.g., aomhdr−>src_addr , aomhdr−>mum_msg_rdr . . .   12
            . . . . . .                                             13
            send ( pkt ,  0);                                       14
            return  (TCL_OK );                                     15
        }                                                          16
    }                                                              17
    . . . . . .                                                    18
```

```
    else if (argc == 6) {                                        19
        if ((strcmp(argv[1], "config_bfsz") == 0)){              20
            src_bfsz_ = atoi(argv[2]);                           21
            tbr_bfsz_ = atoi(argv[3]);                           22
            mum_entry_vol_ = atoi(argv[4]);                      23
            shim_hdr_vol_ = atoi(argv[5]);                       24
            return (TCL_OK);                                     25
        }                                                        26
    }                                                            27
    ......                                                       28
    return (Agent::command(argc, argv));                         29
}                                                                30
```
Program 11. Commands of AOM Agent

6.3.3 Network-Layer AOM Agent

The network-layer AOM agent *AOMAgent* enhances the network router with AOM-aware intelligence. The agent establishes the forwarding states for RPF when MUM messages pass through the local node and performs multicast forwarding when a data packet arrives. Program 12 shows the declaration of *AOMAgent*.

```
// ~/ns/gaon_module/aom_protocol/aomagent.h                     1
#include ~/ns/apps/aom_app.h                                    2
                                                                3
#include ~/ns/gaon_module/RouteClassifier.h                     4
                                                                5
#include bloom_filter.h                                         6
                                                                7
class AOMAgent : public Agent{ public:                          8
    AOMAgent();                                                 9
    enum {SDR, TBR, RDR, LRT}; // LRT, legacy router           10
    int rtr_type_;    // Type of local router, e.g., RDR       11
    int tbr_bfsz_;    // IRDR_BF size                          12
    string ip_prefix_;   // IP address of local AOMAgent       13
    BFLst irdr_bfs_;   // IRDR_BFs at this TBR's AOMAgent      14
    void JoinMsgTBRProc();    // Process joining msg           15
    void FastJoinTBRProc();    // Process fast-join msg        16
    void LeaveMsgTBRProc();    // Process leaving msg          17
    void DataFwdingTBRProc();    // Forward data pkts          18
    void DataFwdingLRTProc();    // Legacy rtr forward pkts    19
    virtual int command(int argc, const char*const* argv);     20
    virtual void recv(Packet*, Handler*);                      21
    ......                                                     22
};                                                             23
```
Program 12. Declaration of AOMAgent

The types of each *AOMAgent* object is configured when creating the simulation topology through OTCL command. The types of the router can be any from the list in line 8 of Program 12. In order to install forwarding states in corresponding interface, the bloom filter size of the IRDR_BF should be specified (line 10). The ns-2 uses a simplified addressing scheme. The attribute *ip_prefix_* (line 11) is for assigning the 32-bit IP addresses to the router. Multicast forwarding states are contained in *irdr_bfs_* (line 12). When a packet arrives at the agent, function *recv()* (line 19) will be executed. Within *recv()*, functions are further invoked to process different types of messages. For a joining message, the forwarding states will be installed by *JoinMsgTBRProc()* (line 13), while the *DataFwdingTBRProc()* is invoked to perform AOM forwarding protocol when receiving a multicasting data packet (line 16). In order to implement fast group joining, leaving, and incremental deployment, the agent implements these operations using corresponding functions (lines 14, 15, and 17). These function bodies are defined in file *aomagent.cc*, which are largely pure C++ programming.

Both application and network layer AOM agents make use of bloom filters. The class *bloom_filter* is declared as in Program 13. The constructor of the bloom_filter object requires the size of the bit vector to be specified (line 4). A string key is hashed by *k* (lines 11–15) times to get *k* bit positions. These positions will be recorded in *bf_bitvec* (line 7). Those hash functions are for general purpose, which can be found in [1]. The operation *insert()* is defined in lines 18–23. The number of keys to be inserted into the bloom filter should be tracked to maintain reasonable false-positive rate (line 9).

```
// ~/ns/gaon_module/aom_protocol/bloom_filter.h          1
class bloom_filter{ public:                               2
    bloom_filter(unsigned int bf_size);                   3
    void insert(const std::string& key);                  4
private:                                                  5
    std::set<unsigned int> bf_bitvec;                     6
    unsigned int bf_bitvec_size;                          7
    unsigned int key_count;                               8
    // Hash functions:                                    9
    unsigned int RSHash  (const std::string& str);        10
    unsigned int JSHash  (const std::string& str);        11
    unsigned int PJWHash (const std::string& str);        12
    unsigned int ELFHash (const std::string& str);        13
    unsigned int BKDRHash(const std::string& str);        14
};                                                        15
// ~/ns/gaon_module/aom_protocol/bloom_filter.cc          16
void bloom_filter::insert(const std::string& key){        17
    string str_key;  // Key to be hashed                  18
    unsigned int hash;  // Yielded number after hashing   19
    key_count++;  // Record how many elements in the BF.  20
    hash = RSHash(str_key)% (table_size);                 21
                        // Hashing with RSHash             22
    bf_bitvec.insert(hash);                               23
    ......                                                24
}                                                         25
```

Program 13. Bloom filter class

The implementation of AOM agent in C++ domain is accomplished so far. In order to make AOM agents effect, all the newly created files should be added to the make file of ns-2 before recompilation. Detailed instructions on how to integrate new codes into ns-2 can be found in [3].

6.4 Summary

In this chapter, we have presented an ns-2-compatible GAON simulation framework. GAON developers could conveniently enhance the ns-2 node with developer-defined GAON agents, which realizes the incorporation of experimental application-oriented intelligence into network nodes, without worrying about the compatibility with regular ns-2 utilities. With GAON, the user can flexibly load/unload specific GAON agents on the ns-2 node through a generic scenario control interface. The ns-2 node structure is extended to properly accommodate GAON components, where there is a GAON agent classifier dispatching GAON traffic to corresponding developer-defined agents while the regular unicast traffic is not affected. Moreover, a unified interface is developed for GAON agents to have access to the ns-2 built-in routing table for packet forwarding. With its simple and effective architecture, GAON provides explicit interfaces to powerful but veiled ns-2 functionalities without introducing any other software technique, which facilitates the evolution of GAON with ns-2 package updating. We have shown the key implementations of AOM protocol under GAON.

References

1. General purpose hash function algorithms. http://www.partow.net/programming/hashfunctions
2. Hwang W, Ke C, Shieh C, Ziviani A (2008) An evaluation framework for more realistic simulations of MPEG video transmission. J Inf Sci Eng 24:425–440
3. Implementing a new nanet unicast routing protocol in ns2. http://masimum.inf.um.es/nsrt-howto/pdf/nsrt-howto.pdf
4. Issariyakul T, Hossain E (2009) Introduction to network simulator ns2. Springer, New York
5. Ousterhout JK (1990) Tcl: an embeddable command language. In: Proc. USENIX, 1990, pp 133–146
6. Tian X, Cheng Y, Liu B (2011) A generic application-oriented networking (GAON) simulation framework for next-generation internet. In: Proc. IEEE ICC
7. Unified modeling language. http://en.wikipedia.org/wiki/Unified_Modeling_Language

Chapter 7
Conclusions and Open Research Issues

Abstract This chapter gives conclusion remarks and possible open research issues.

7.1 Conclusions

This book has proposed a scalable and efficient multicast mechanism with the approach of application-oriented networking (AON), the essence of which is to integrate application-level intelligence into the network. The AON-based multicast proposed is formally termed as application-oriented multicast (AOM). AOM eliminates the need to maintain group-specific membership information at routers along the multicasting tree, thus yields desirable scalability compared with IP multicast; moreover, the bandwidth overhead of AOM is close to that of IP multicast, which is achieved by the proposed bloom filter-based design. Specifically, the main contribution of the research is summarized as the following:

- *Scalability Improvement.* In the AOM service model, the packet carries the explicit destination addresses in its header, instead of an implicit multicast IP address, to facilitate the multicast forwarding; routers with application-level intelligence can retrieve the addresses and leverage the local forwarding states to determine necessary packet copies and the corresponding forwarding interface for each copy. In the bloom filter-based design, the forwarding states at each router are destination specific, only depending on the number of receivers that can be reached through the router. The forwarding complexity is totally independent of the number of groups being supported by the router. Compared with IP multicast, AOM installs less forwarding states at routers on the multicasting tree. This is because AOM stores only one state on each related node for each subscribing domain. In contrast, each subscribing domain may join in tens of thousands of groups, and each group needs a state on each related node under IP multicast. Moreover, group IDs in AOM are only used for labeling groups at the data source and subscribing domains to establish the service relationship;

X. Tian and Y. Cheng, *Scalable Multicasting over Next-Generation Internet: Design,* 147
Analysis and Applications, DOI 10.1007/978-1-4614-0152-0_7,
© Springer Science+Business Media New York 2013

thus, the group IDs can be allocated at the data source locally in the form of a two-tuple (source node address, source-specific channel ID), which breaks the address space limitation of IPv4 Class-D addresses.

- *Balance Between Bandwidth Efficiency and false-positive.* The bandwidth over-head of the explicit addressing in AOM is constrained by the proposed bloom filter-based design. The destination addresses in the packet are encoded in the format of a bloom filter. AOM strikes a balance between bandwidth efficiency and small bloom filter false-positive rate, in comparison with the most closely related work FRM, another multicast protocol-based on bloom filter. The funda-mental difference between AOM and FRM is that AOM encodes only destination prefixes in the packet's shim header while the FRM encodes multicasting tree branches. The bloom filter incurs false-positive, which means the element not encoded into the bloom filter might be falsely detected. For a fixed-length bloom filter, the more elements are encoded, the higher the false-positive rate can be. In AOM and FRM, when the number of receivers/branches exceeds the capacity of a single shim header, multiple packets are sent to cover all destinations, which are counted as redundant traffic. Since covering the same number of destinations normally requires more branches, AOM can generate less redundant traffic than FRM does if they keep the same false-positive rate.

- *Practical Design.* Some important practical design issues of the bloom filter-based multicast protocol are solved with efficiency. A BGP-view-based group joining process is proposed to deal with the asymmetric interdomain routing issue, which avoids the complexity of deploying/configuring the MBGP protocol and enables a fast group joining mechanism as a side-benefit. The fast-join scheme can shorten the joining delay perceived by receivers under AOM, com-pared with FRM. Moreover, the bloom filter performance of AOM is evaluated with numerical and theoretical approaches. Compared with other bloom filter-based multicast protocols, AOM can automatically eliminate the forwarding loop caused by the bloom filter false-positive, without any assistance from underlying infrastructure. Further, an incremental deployment of AOM is proposed. AOM can be deployed in a network, in which only a small fraction of AOM-aware routers exist while others are legacy routers.

- *AOM-Assisted IPTV Channel Zapping Acceleration.* An AOM-assisted channel zapping acceleration (AAZA) scheme has been proposed to IPTV systems to reduce the IPTV channel zapping time, which is an important QoE metric in IPTV service. The recent zapping acceleration scheme based on time-shifted subchannels (TSS) is systematically analyzed from a theoretical perspective. The analysis shows that there exists an optimal subchannel data rate which minimizes the redundant traffic transmitted over subchannels; moreover, the *start-up effect* is revealed, where the original operation pattern could violate the performance bound of the TSS-based model. With a convenient solution to the start-up effect proposed, a rigorous proof is presented, which shows that the bounded zapping time is guaranteed in the TSS-based model. Furthermore, the design of AAZA scheme is described to implement the TSS-based model with the facilitation of AOM. In contrast to the IP multicast-based implementation, AAZA achieves the seamless subscriber-migration from the subchannel to the main channel, without any control message exchange over the network; the subchannel selection in AAZA is independent of the zapping request signaling time, resulting in improved robustness and reduced messaging overhead. We implement AAZA in ns-2 and multicast an MPEG-4 video stream over a practical network topology. Extensive simulation results are presented to demonstrate the validity of our analysis and AAZA scheme.

- *Generic GAON Framework.* The AON technique introduced in this thesis is one of the mainstream ideas to design the next-generation Internet, explicitly or implicitly. The systematical study of AON needs a flexible and user-friendly simulation tool. This thesis has presented an ns-2-compatible generic application-oriented networking (GAON) simulation framework to facilitate AON research. GAON developers could conveniently enhance the ns-2 node with developer-defined GAON agents, which realizes the incorporation of experimental AON intelligence into network nodes without worrying about the compatibility with regular ns-2 utilities. With GAON, the user can flexibly load/unload specific GAON agents on the ns-2 node through a generic scenario control interface. The ns-2 node structure is extended to properly accommodate GAON components, where there is a GAON agent classifier dispatching GAON traffic to corresponding developer-defined agents while the regular unicast traffic is not affected. Moreover, a unified interface is developed for GAON agents to have access to the ns-2 built-in routing table for packet forwarding. With its simple and effective architecture, GAON provides explicit interfaces to powerful but veiled ns-2 functionalities without introducing any other software technique, which facilitates the evolution of GAON with ns-2 package updating.

7.2 Open Research Issues

There are many open research issues currently remaining to be solved for the next-generation Internet, which can be streamlined and redesigned with the rapid development in the hardware/software technologies. This thesis initiated the studies on this topic, and further research in the following challenging fields for contributing to an application-oriented, information-centric Internet is planned in the future.

- *AON.* This thesis starts to investigate AON technique by leveraging the AON processing in routers to facilitate multicasting; however, continuing efforts are still needed to push the research in the area forward. The proposed AON multicasting is planed to be implemented over the PlanetLab testbed, so that the performance of AON multicast over the practical Internet could be better illustrated and the design could be further optimized. Moreover, some interesting topics including but not limited to the following will be examined: whether other common networking functionalities such as Anycast could benefit from the AON technology; how the AON router could help the content-based routing over the Internet, compared to over the enterprise-wide networks; AON concept is one of the mainstream ideas to design the next-generation Internet.

- *Video-on-Demand (VoD) Service Assisted by AOM.* The AAZA scheme described in the thesis focuses on the broadcast TV channel service, while another scenario of IPTV is VoD. In the broadcasting scenario, users may miss some portions of the TV program. They could watch the ongoing program after their requests have been processed. However, every user in the VoD scenario is supposed to watch from the beginning of the video. The naive solution is to make some requests wait for a while and aggregate following requests whose arrival times are within some specific range. Then the VoD server can multicast one stream to serve them. However, long waiting time affects the QoE of end users while short waiting time can make the VoD server overloaded. Intuitively, the AAZA scheme can be modified to deliver VoD service because the TSS-based model in nature aggregates requests. Recall that subscribers to different subchannels will eventually migrate to the main channel. Nevertheless, as the arrival times of the requests may have a large variance and each user wants to watch the video from the beginning, the redundant traffic generated by subchannels could be significant. The preliminary idea is to partition the object into segments in order to reduce the redundant traffic. The future work will elaborate how to modify the AAZA scheme to provide VoD service.

- *P2P Networking Over AON Networks.* P2P systems have been widely utilized in the current Internet. Traditionally, P2P systems are categorized into structured and unstructured. Structured P2P systems based on dynamic hash table (DHT) have the content knowledge, which associates requested content with the peer owning the content through the DHT. The structured P2P systems could cause long request latency, as the topology knowledge about the underlying transport network is unknown. Unstructured P2P systems have topology knowledge, where each peer maintains the information about some of its neighbors physically connected to it. However, the lack of content knowledge makes unstructured overlay network hard to scale, since it incurs considerable redundant traffic. Intuitively, the performance of a P2P system could be improved if both content knowledge and topology knowledge could be obtained. This complies with the concept of AON, where the underlying infrastructure could provide more application-level services. The application of network coding to P2P system is an example of such philosophy. Implementing P2P systems over AON networks could be an interesting topic to obtain some insight into the impact of the AON technology on popular applications.

- *Cloud Computing.* Cloud computing is a recent trend in IT service paradigm, which moves computing and data from local computers to remote data centers (DCs) located in various parts of the network. It is another example hinting at a future in which the Internet will become a data-centric infrastructure. DCs are interconnected computer networks, where all the components, e.g., telecommunications and storage, are controlled by a single entity. A high-performance and robust networking architecture could greatly facilitate the intra-/inter-DC communications centralized around data, where the bandwidth-intensive one-to-one, one-to-several, and all-to-all communication paradigms are efficiently implemented. In this sense, the concept of AON could be easily adapted to cloud computing. For example, the interaction nature between computing resources and the data to be processed requires the network to deliver data to the close-to computing power in order to reduce data movements and improve scalability. This task will be better served by a network with combined intelligence of computing power and data storage, which is within the scope of AON networking technique. Furthermore, it is only a matter of time for the cloud to scale up; if there are multiple clouds over the Internet, it is possible for a cloud to "borrow" the computing power from other peering clouds. Dispatching data based on the kind of resources needed and finding the clouds where the resources are available are similar to the principle of the service-oriented architecture (SOA), which could be facilitated by a more powerful AON networking fabric.

Index

X. Tian and Y. Cheng, *Scalable Multicasting over Next-Generation Internet: Design,*
Analysis and Applications, DOI 10.1007/978-1-4614-0152-0,
© Springer Science+Business Media New York 2013